Winemaking: Fermenting, Pressing, Bottling, and Aging

An Introduction to Oenology

Contents

1 Oenologists   **1**

1.1 Education and training 1

     1.1.1 Australia 1

     1.1.2 Canada 1

     1.1.3 France 1

     1.1.4 Italy 1

     1.1.5 New Zealand 2

     1.1.6 Portugal 2

     1.1.7 South Africa 2

     1.1.8 Switzerland 2

     1.1.9 Spain 2

     1.1.10 United Kingdom 2

     1.1.11 United States 2

1.2 See also 2

1.3 References 2

1.4 External links 2

2 Maynard Amerine   **3**

2.1 Selected publications 3

2.2 References 3

2.3 External links 3

3 Christine Barbe   **4**

3.1 Personal life and education 4

3.2 Career 4

     3.2.1 Toquade Wines 4

3.3 References 4

3.4 External links 4

4 Ruy Barbosa Popolizio   **5**

4.1 References 5

5 Valentin Blattner **6**

5.1 Varieties created or propagated 6

5.2 References 6

6 Yiannis Boutaris **8**

6.1 Early life 8

6.2 References 8

6.3 External links 8

7 Cathy Corison **9**

7.1 Personal life and education 9

7.2 Career 9

7.2.1 Corison Winery 9

7.2.2 Awards and acknowledgements 9

7.3 References 9

7.4 External links 9

8 Tullio De Rosa **11**

8.1 Career 11

8.2 References 11

9 Sarah Gott **12**

9.1 Personal life and education 12

9.2 Career 12

9.3 References 12

9.4 External links 12

10 Belinda Kemp **13**

10.1 Career 13

10.2 References 13

11 Mia Klein **14**

11.1 Personal life and education 14

11.2 Career 14

11.2.1 Selene Wines 14

11.3 References 14

11.4 External links 15

12 Max Léglise **16**

12.1 Selected publications 16

12.2 See also 16

13 Zelma Long   **17**

13.1 Career 17

13.2 References 17

14 Justin Meyer   **19**

14.1 Background 19

14.1.1 Career 19

14.2 Death and legacy 20

14.3 References 21

15 Hermann Müller (Thurgau)   **23**

15.1 Biography 23

15.2 Breeding of the Müller-Thurgau grape variety 24

15.3 External links 24

15.4 References 24

16 List of oenologists   **25**

16.1 Oenologists 25

16.2 See also 25

17 Ottavio Ottavi   **26**

17.1 Biography 26

17.2 References 26

17.3 External links 26

18 Émile Peynaud   **27**

18.1 Biography 27

18.2 Selected bibliography 27

18.3 See also 27

18.4 References 27

19 Jacques Puisais   **29**

19.1 See also 29

19.2 References 29

20 Michel Rolland   **30**

20.1 Education and early career 30

20.2 Media exposure 31

20.3 Influence 31

20.4 Bordeaux vineyards under Rolland influence 31

20.5 See also 32

20.6 Sources 32

21 Carol Shelton   **33**

21.1 Personal life and education 33

21.2 Career 33

     21.2.1 Carol Shelton Wines 33

     21.2.2 Awards 33

21.3 References 34

21.4 External links 34

22 Kilien Stengel   **35**

22.1 Works 35

     22.1.1 Actuality books 35

     22.1.2 History books 35

     22.1.3 Poésie books 36

     22.1.4 Trivia books 36

     22.1.5 Games 36

     22.1.6 Practical books 36

     22.1.7 School books 36

     22.1.8 Part of book and direction of book 36

     22.1.9 Collaborations reviews 37

     22.1.10 Awards 37

22.2 References 37

23 Lane Tanner   **38**

23.1 Personal life and education 38

23.2 Career 38

     23.2.1 Lane Tanner Winery 38

23.3 References 38

23.4 External links 39

24 Păstorel Teodoreanu   **40**

24.1 Biography 40

     24.1.1 Early life and World War I service 40

     24.1.2 Debut and *Gândirea* affiliation 40

     24.1.3 *Țara Noastră* and *Rodia de aur* 41

     24.1.4 1930s 42

      24.1.5   World War II propagandist 43

      24.1.6   Communist takeover 43

      24.1.7   Censorship and show trial 44

      24.1.8   Illness and death 44

  24.2 Work 45

      24.2.1   Jester Harrow 45

      24.2.2   Caragialesque prose 46

      24.2.3   Symbolist poetry 47

      24.2.4   Scattered texts and apocrypha 47

  24.3 In cultural memory 47

  24.4 Notes 48

  24.5 References 51

  24.6 External links 51

25 Keith Wallace (wine writer)   **52**

  25.1 Controversy 52

  25.2 References 52

  25.3 External links 52

  25.4 Text and image sources, contributors, and licenses 53

      25.4.1   Text 53

      25.4.2   Images 54

      25.4.3   Content license 55

Chapter 1

Oenologists

Oenology (**enology** in American English; pronounced /iːˈnɒlədʒi/ *ee-NOL-o-jee*) is the science and study of all aspects of wine and winemaking except vine-growing and grape-harvesting, which form a subfield called viticulture. "Viticulture and oenology" is a common designation for training programmes and research centres that include both the "outdoors" and "indoors" aspects of wine production. An expert in the field of oenology is known as an **oenologist**. The word **oenology** is derived from the Ancient Greek word οἶνος (oinos, "wine") and the suffix -λογία (-logia, "study of").

1.1 Education and training

An increasing number of schools offer degree-granting programs in Oenology and Viticulture. Most of these offer it as a major concentration for a Bachelor of Science (B.S, B.Sc., Sc.B) degree or offer it as a terminal master's degree—either in scientific or research oriented program culminating in a Master of Science (M.S., Sc.M.) degree, or a professional degree, like Cornell University's Master of Professional Services. Oenologists and viticulturalists who hold doctoral degrees often have earned them in related fields, including Horticulture, plant physiology or microbiology. Related non-academic titles including sommelier and Master of Wine certifications are more oriented toward commercial occupations in the restaurant or hospitality management industry.

Oenologists often work as winemakers or find employment with commercial laboratories or with groups such as the Australian Wine Research Institute.

1.1.1 Australia

Schools in Australia tend to offer a "Bachelor of Viticulture" or "Masters of Viticulture" degree.

- Charles Sturt University - Wagga Wagga, New South Wales

- Curtin University of Technology - Perth, Western Australia

- Northern Melbourne Institute of Technology - Preston, Victoria

- University of Adelaide - Adelaide, South Australia

- Queensland College of Wine Tourism - Stanthorpe, Queensland

1.1.2 Canada

- Brock University - St. Catharines, Ontario

1.1.3 France

Official *National Diploma of Oenology* :

- Jules Guyot Institute - Dijon

- University of Montpellier I - Montpellier

- Institut National Polytechnique de Toulouse - Toulouse

- University of Reims - Reims

- University of Bordeaux - Bordeaux

Other wine diplomas :

- Université du Vin - Suze-la-Rousse

- Université de Bourgogne - Dijon

1.1.4 Italy

- University of Padua - Padua

1.1.5   New Zealand

- Lincoln University - Christchurch
- Eastern Institute of Technology - Hawke's Bay

1.1.6   Portugal

- Universidade do Porto - Porto
- University of Trás-os-Montes and Alto Douro - Vila Real

1.1.7   South Africa

- University of Stellenbosch - Stellenbosch

1.1.8   Switzerland

- Changins - Nyon

1.1.9   Spain

· Universitat Rovira i Virgili - Tarragona · Universidad de La Rioja- La Rioja

1.1.10   United Kingdom

- Blake Hall College - London
- Plumpton College - East Sussex[1]

1.1.11   United States

- Cornell University - Ithaca and Geneva, New York
- Paul Smith's College - Paul Smiths, New York
- Finger Lakes Community College - Canandaigua and Geneva, New York
- California Polytechnic State University - San Luis Obispo, California
- Fresno State University - Fresno, California
- Sonoma State University - Sonoma, California
- University of California, Davis - Davis, California
- Allan Hancock College - Santa Maria, California
- Napa Valley College - Napa, California
- Virginia Polytechnic Institute and State University (Virginia Tech) - Blacksburg, Virginia
- Washington State University - Pullman, Washington
- Patrick Henry Community College - Henry County, Virginia
- Surry Community College - Dobson, North Carolina
- Grayson College - Grayson, Texas
- University of Missouri - Columbia, Missouri
- Kent State - Kent, Ohio
- Miami University - Oxford, Ohio
- Colorado State University - Fort Collins, Colorado
- Yavapai College - Clarkdale, Arizona
- Oregon State University - Corvallis, Oregon

1.2   See also

- List of oenologists
- Master of Wine
- Sommelier

1.3   References

[1] "Bachelor Degree (BSc Hons) Viticulture and Oenology". Plumpton College. Retrieved 21 July 2014.

1.4   External links

- Bibliography of basic current books about wine and winemaking
- Oenologist.com - Online resource about wine and professionals in the wine industry
- VinoEnology.com - Online wine-making calculators and resource about wine and professionals in the wine industry
- Glossary of Terms for Enology, Viticulture and Winemaking, CCwinegroup.com, 2009 (PDF)

Chapter 2

Maynard Amerine

Maynard Amerine (1911–1998) was a pioneering researcher in the cultivation, fermentation, and sensory evaluation of wine. His academic work at the University of California at Davis is recognized internationally. His 16 books and some 400 articles contributed significantly to the development of the modern (post-Prohibition) wine industry in California; to the improvement of wine cultures in Europe, South America, and Australia; and to the professional standards for judging and tasting wine.

In the early 1940s, he and his colleague Albert J. Winkler developed the Winkler scale, a technique for classifying wine growing regions based on temperatures, that continues to be used in the United States and elsewhere. His research, organizational, and advisory efforts in wine tasting helped bring about a more objective vocabulary to that field, based on flavors and scents rather than allusive references.[1]

Maynard Andrew Amerine was born in 1911 in San Jose, California, the child of Roy Reagan Amerine and Tennessee Davis Amerine. He grew up on their farm in Modesto, California. In 1935, while still completing his Ph.D. in plant physiology at the University of California at Berkeley, he became the first faculty member hired into the new Viticulture and Enology Department at the University of California at Davis. He became full professor in 1952, was deemed All-University Lecturer in 1962, and continued his research and teaching at Davis until his retirement in 1974. He remained Professor Emeritus, served as an advisor at the Wine Institute in San Francisco, and continued to write, travel, and advise on wine production and evaluation nearly until his death in 1998.

The first named professorship at the University of California was endowed in his honor as the Maynard Amerine Chair, by Ernest Gallo (an early classmate of Amerine's). In 1991, the Maynard A. Amerine Room and Wine Collection in Shields Library was named in his honor and to house his extensive collection of books, a product of his years as a wine bibliographer and collector.

2.1 Selected publications

- 1965. *Wine, An Introduction.* Revised edition 1975 with Vernon L. Singleton.

- 1976. *Wines: Their Sensory Evaluation,* with Edward B. Roessler (W.H. Freeman & Company). Revised and enlarged, 1983.

- *Table Wines and Dessert and Appetizer Wines,* with Maynard A. Joslyn.

- *Technology of Winemaking,* with William V. Cruess, Harold W. Berg; revised with Ralph E. Kunkee, Cornelius S. Ough, Vernon L. Singleton, and A. Dinsmore Webb.

2.2 References

[1] Lawrence Osborne, *The Accidental Connoisseur: An Irreverent Journey Through the Wine World* (2004), pp. 130-131.

- "Maynard A. Amerine, Viticulture and Enology: Davis," in *University of California: In Memoriam, 1998, editid by David Krogh (Oakland, CA: Academic Senate, University of California, 1998).*

- Obituary: "Maynard Amerine, 87, California Wine Expert," New York Times, 13 March 1998.

2.3 External links

- Maynard Amerine Papers at Special Collections Dept., University Library, University of California, Davis

- Video of class taught by Amerine: Viticulture and Enology 125: Sensory Analysis of Wine, from the UCD Library Special Collections

- Biography of Amerine, from the UCD Library

Chapter 3

Christine Barbe

Christine Barbe is a French winemaker. Sauvignon blanc is her specialty.

3.1 Personal life and education

Christine Barbe was born in Bordeaux, France. She attended the Bordeaux Institute of Enology, where he received her Ph.D. in 1996 in Enology and Viticulture. She relocated, after graduation, to California in the United States.[1]

3.2 Career

Barbe first started working in the wine industry in 1991 in France, while getting her Ph.D. She worked at Château Carbonnieux and Château La Louvière.[1] Denis Dubourdieu was a mentor for Barbe, who taught her about Sauvignon blanc wine.[2] Upon relocating to California, she was hired by E & J Gallo Winery.[3] She also worked at Robert Mondavi and Trinchero Family Estates. In 2006, she became winemaker for Cockerell Family Wine Estates. In 2009, she took on a larger management role at the company, but still retaining her winemaker title.[1]

3.2.1 Toquade Wines

Barbe's own wine label, Toquade Wines, makes Sauvignon blanc wine. The word *toquade,* means "infatuation" or "craze" in French. She named the label Toquade as a tribute to her early experiences in the wine industry in France.[1] Toquade's first vintage was in 2006 and consisted of 100 cases. The 2007 vintage produced 200 cases. The 2008 wine was made from fruit from a dry farmed single block vineyard in Yountville, California. The wine was fermented and aged in stainless steel tanks. It was aged for seven months.[3] The 2010 production produced 450 cases.[4]

3.3 References

[1] "Christine Barbe". *Winemakers A-Z.* Women winemakers of California. Retrieved 22 December 2012.

[2] Seda, Catherine (7 June 2012). "Wine of the Week: Terroir Coquerel Sauvignon blanc 2009 Napa Valley". *St. Helena Star.* Retrieved 22 December 2012.

[3] "Toquade Wines". *Napa Valley Wineries.* The Napa Wine Project. Retrieved 22 December 2012.

[4] Teague, Lettie. "Sampling Sauvignon blanc". *On Wine.* The Wall Street Journal. Retrieved 22 December 2012.

3.4 External links

- Official website for Toquade Wines

- Interview with Christine Barbe

Chapter 4

Ruy Barbosa Popolizio

Ruy Barbosa Popolizio (December 2, 1919 – June 8, 2014) was a Chilean businessman, politician, and oenologist. Barbosa served as the Minister of Agriculture from September 26, 1963, until November 3, 1964, as well as the Minister of National Assets briefly from September to November 1963. He later became the Rector of the University of Chile from 1968 until 1969.[1]

Barbosa Popolizio died in Santiago, Chile, on June 8, 2014, at the age of 94.[1]

4.1 References

[1] "La Universidad de Chile despide al ex rector Ruy Barbosa". *University of Chile*. 2014-06-09. Retrieved 2014-07-06.

Chapter 5

Valentin Blattner

Valentin Blattner is a Swiss grape geneticist, grape breeder and winemaker of the Jura Mountains.[1] Blattner has conducted very important research into finding disease-resistant grapes in viniculture since the 1980s,[2] and is best known for developing Cabernet blanc in his Soyhières nursery in 1991.[3] He crossed varieties of *vinifera* with other subspecies, which have since become known as "Blattners".[2] In making his wines, he relies on traditional field breeding techniques. He has a position at the Institute of Ecology and Grape Breeding in Switzerland.[4]

5.1 Varieties created or propagated

- Birstaler Muskat (a Bacchus x Seyval blanc crossing)[5]

- Cabernet blanc (a Cabernet Sauvignon x Resistenzpartner crossing)[6]

- Cabernet Jura (a red grape crossing of Cabernet Sauvignon x Resistenzpartner)[7]

- Cabernet noir (a crossing of Cabernet Sauvignon with an unknown variety)[8]

- Cabertin (another Cabernet Sauvignon x Resistenzpartner crossing)[9]

- Pinotin (a Cabernet Sauvignon x Resistenzpartner crossing)[10]

- Reselle (another Bacchus x Seyval blanc crossing)[11]

- Petite Milo (unknown x Resistenzpartner crossing) grown in British Columbia and Nova Scotia, Canada

- Epicure (a Cabernet Sauvignon x Resistenzpartner crossing) grown in British Columbia, Canada named after Victorian Epicure a company in Victoria BC that lent much support in the development of the Blattner crosses in Canada

- Cabernet Foch (Cabernet Sauvignon x Foch crossing) grown in British Columbia and Nova Scotia, Canada

- Amiel (Cabernet Sauvignon x Resistenzpartner crossing) grown in British Columbia, Canada. Early ripening white grape ripening about the same time as Ortega in the BC coastal climate.

- Labelle (Cabernet Foch x Resistenzpartner crossing) grown in British Columbia, Canada. An very early ripening red grape.

5.2 References

[1] "Blattner history". Omegavines.com. Retrieved 3 June 2013.

[2] Hynes, Gary (March 2011). *Island Wineries of British Columbia*. TouchWood Editions. p. 125. ISBN 978-1-926741-26-0. Retrieved 3 June 2013.

[3] Pitte, Jean-Robert (2010). *Il desiderio del vino. Storia di una passione antica* (in Italian). EDIZIONI DEDALO. p. 31. ISBN 978-88-220-0572-4. Retrieved 3 June 2013.

[4] *Wine East*. L&H Photojournalism. 2006. p. 4. Retrieved 3 June 2013.

[5] Vitis International Variety Catalogue (VIVC) *Birstaler Muskat* Accessed: June 3rd, 2013

[6] Vitis International Variety Catalogue (VIVC) *Cabernet blanc* Accessed: May 27th, 2013

[7] Vitis International Variety Catalogue (VIVC) *Cabernet Jura* Accessed: June 3rd, 2013

[8] Vitis International Variety Catalogue (VIVC) *Cabernet noir* Accessed: June 3rd, 2013

[9] Vitis International Variety Catalogue (VIVC) *Cabertin* Accessed: June 3rd, 2013

[10] Vitis International Variety Catalogue (VIVC) *Pinotin* Accessed: June 3rd, 2013

[11] Vitis International Variety Catalogue (VIVC) *Reselle* Accessed: June 3rd, 2013

Chapter 6

Yiannis Boutaris

Yiannis Boutaris (Greek: Γιάννης Μπουτάρης; born 13 June 1942[1]) is a Greek businessman, politician and current mayor of Thessaloniki. In 2012 he was chosen as 'the best mayor of the world' for the month of October, by the City Mayors Foundation, based in the UK. He is the founder of *KIR-YIANNI* wine company, based in Giannakochori and Amyntaio. He is one of the founding members of the Drasi party and Arcturos ecological organization.[2]

6.1 Early life

Yiannis Boutaris was born in Thessaloniki in 1942, the son of winemaker Stelios Boutaris and Fanny Vlachos, from the Vlach *Nichota* family in the town of Krusevo, Republic of Macedonia.[3]

His primary education was at the Experimental elementary school of the Aristotle University of Thessaloniki, his secondary education at Anatolia College, and he graduated in chemistry from the Aristotle University of Thessaloniki in 1965 and in oenology from the Wine Institute of Athens in 1967.[1][4] In his youth he was associated with the Communist Party of Greece (KKE).[5]

From 1969 to 1996 he worked for the family wine company Boutari, based in Naoussa. He left the family company to create the Kir-Yianni wine company, based on two estates in Giannakochori and Amyntaio, in 1998.

6.2 References

[1] "Yannis Boutaris speaker profile". *IMIC 2012 conference.* 15–16 February 2012. Retrieved 2013-11-05.

[2] "Γιάννης Μπουτάρης" (in Greek). Drasi. Retrieved 15 November 2010.

[3] "The Mayor of Thessaloniki Yiannis Boutaris will build a house in his hometown Krusevo, identical to the one where his mother, from the Vlach Nichota family, lived, with rec-ognizable ambiance and traditional Krusevo architecture.". *Kanal 5* (in Macedonian).

[4] Γιάννης Μπουτάρης Βιογραφικό σημείωμα υποψηφίου Δημάρχου Θεσσαλονίκης (in Greek). protovoulia2010.gr. Retrieved 15 November 2010.

[5] "Thessaloniki mayor, Golden Dawn clash over bear sanctuary donation". *Kathimerini.* 22 August 2013. Retrieved 2013-11-05.

6.3 External links

- City Mayors' Mayor of the Month for October 2012
- Official campaign website
- KIR-YIANNI company website
- *Local elections '10: Vintage Boutaris for Thessaloniki,* article by George Gilson in *Athens News,* 13 September 2010

Movie "One Step Ahead" (2012) "Ena vima brosta" (original title) Directed by Dimitris Athyridis in 2012.

Chapter 7

Cathy Corison

Cathy Corison is an American winemaker and consultant. She specializes in Cabernet Sauvignon. Corison was the *San Francisco Chronicle* Winemaker of the Year in 2011.

7.1 Personal life and education

Cathy Corison grew up in Riverside, California.[1] She studied biology at Pomona College and was on their men's diving team, because the school didn't have a women's team.[1][2] In 1972, she had to take an extracurricular class. Using her diving skills she decided to sign up for a trampoline class, but changed her mind upon seeing a sign-up sheet for a wine tasting class.[1] This class was the catalyst for garnering Corison's interest in winemaking. After graduation in 1975, she moved to Napa Valley in California.[1][3] She received her Master's degree in Enology from University of California, Davis.[1][2]

7.2 Career

Upon moving to Napa, she started working in the tasting room at Sterling Vineyards and at a wine shop. During this time, she was getting her Master's degree at the University of California, Davis. She was told by her professor that she would not get a job in Napa Valley because of being a woman. She tried to get a job at Freemark Abbey and was denied the job. Part of the reasons the owners gave was because she was too short and they believed she could not work in the wine cellar. She almost started working at Christian Brothers in the enology lab. Many women worked in labs and weren't allowed to be winemakers. Corison decided against working in the lab and in 1978 she became an intern at Freemark Abbey. She eventually became a winemaker for the winery.[1] She then joined Chappellet Winery, in 1983, as head winemaker, working there for almost ten years.[3][1] At Chappellet she was able to focus on developing her skills at creating Cabernet Sauvignon wines. She dislikes wines with high alcohol content and fruit forward wines.[3] She does not add additives of acids to her wine.[1]

7.2.1 Corison Winery

She founded her own winery, Corison Winery, in 1987. The winery is located in St. Helena, California in a barn built by Corison's husband, William Martin.[1][3] Corison Winery makes Cabernet's and Gewurztraminer wines and produces 3,500 cases a year. The winey makes a Kronos Vineyard Cabernet Sauvignon and a Napa Valley Cabernet Sauvignon. The Kronos is an estate wine and made from organic grapes.[3] The vineyard is dry farmed.[1] The grapes come from one of the oldest vineyards in Napa Valley.[3] The Napa Valley Cabernet comes from Rutherford.[1] The Gewurztraminer is called Corazon and comes from the Anderson Valley AVA.[3]

7.2.2 Awards and acknowledgements

- 2011, *San Francisco Chronicle* Winemaker of the Year[1]

7.3 References

[1] Bonné, Jon (1 January 2012). "Cathy Corison — Chronicle's Winemaker of the Year 2011". *Food* (San Francisco Chronicle). Retrieved 8 December 2012.

[2] Baiocchi, Talia. "Cathy Corison: The Female Clark Kent of Napa Cabernet". *Vintage America*. Eater. Retrieved 8 December 2012.

[3] "Cathy Corison". *People and institutions*. Calwineries. Retrieved 8 December 2012.

7.4 External links

- Cathy Corison on Twitter

- Official website

Chapter 8

Tullio De Rosa

Tullio De Rosa (1923–1994) was an Italian enologist.[1]

8.1 Career

After graduation in Bologna in 1947, De Rosa started teaching in 1966 at the Istituto Sperimentale di Enologia in Conegliano, (formerly part of the Scuola enologica di Conegliano), that he lately directed for several years, the most relevant titles among his wide bibliography include the classic handbooks *Tecnologia dei Vini Bianchi* (White Wines Production Technology), *Tecnologia dei Vini Spumanti* (Sparkling Wines Production Technology), *Tecnologia dei Vini Rossi* (Red Wines Production Technology) and *Tuttovini* (translated in several languages), and the collection of autobiographical novels *Andar Per Vini* that in the 1970 first edition was accompanied by illustrations by the Italian artist Renato Varese. In 2011 the posthumous book *Guida alla degustazione del vino: la valutazione edonistica. Concetti propedeutici e formativi esposti in maniera utilizzabile da un ampio ambito di lettori* (originally manuscripted by De Rosa during his last days of illness between June and August 1994) was finally published jointly by the Microbiogical Institute in Rauscedo and Faenza Editore. He is also the grandfather of the Italian songwriter Fabio Fantuzzi.

8.2 References

[1] Tullio De Rosa (1923-1994) | Università degli Studi di Padova Mar 21, 2013 – this profile is held by the daughter and grandson of prof De Rosa, who died in 1994."

Chapter 9

Sarah Gott

Sarah Gott is an American winemaker.

9.1 Personal life and education

Gott originally attended the University of California, Davis with the intention of studying veterinary medicine. With a long time love for food and wine, she opted to study in enology instead at the University. She received her Bachelor of Science in fermentation science in 1993.[1][2] She is married to winemaker Joel Gott. They have three children. She enjoys participating in triathlons.[3]

9.2 Career

After graduation, Gott interned at Joseph Phelps Winery located in St. Helena, California, Gloria Ferrer in Sonoma, California, and Wirra Wirra in Australia. She completed these internships in three years.[1] While at Joseph Phelps, she was mentored by Craig Williams and worked in the winery labs as an assistant enologist.[1][2] She became assistant winemaker at Joseph Phelps in 1994.[1][3] In 2002, she became head winemaker. In 2004, she started working as the first full-time winemaker at Quintessa Wine.[1][2] Gott left Phelps, where she was responsible for producing 90,000 cases of various wines each year, to create a small batch of wine at Quintessa, totaling 10,000 cases.[2]

Two years later, in 2004, she started working as winemaker at Oakville East Wine Company.[1] She has provided consulting services for Clif Family Winery and Blackbird Vineyards.[3][4] The first Merlot she created for Blackbird Vineyards, a 2003 Merlot, was awarded 95-points from Vintrust.[5] Today, Gott serves as Director of Winemaking at Joel Gott Wines, which she co-runs with her husband, winemaker Joel Gott.[1][6]

9.3 References

[1] Gilbert, Lucia and John. "Sarah Gott". *Winemakers*. Women Winemakers of California. Retrieved 21 December 2012.

[2] Boyd, Gerald D. (29 September 2002). "Winemakers to Watch: Sarah Gott". *Boyd*. Retrieved 21 December 2012.

[3] Boone, Virginia. "A Touch of Advice". *General articles*. North Bay Biz. Retrieved 21 December 2012.

[4] "Sarah Gott is Clif Bar winemaker". *Wines & Vines*. 1 December 2007.(subscription required)

[5] Anash, Sao. "Magical Merlot". *Wine*. Santa Barbara Independent. Retrieved 21 December 2012.

[6] "Bios". *Trade*. Joel Gott Wines. Retrieved 21 December 2012.

9.4 External links

- Official website

Chapter 10

Belinda Kemp

Belinda Sarah Kemp is a British enologist.

10.1 Career

After graduation in Plumpton College with a First Class Bachelor of Science (hons) in Viticulture and Oenology, Kemp completed her PhD thesis on the effects of vine leaf removal on fruit ripening at Lincoln University (New Zealand) beginning in April 2007 and finishing in August 2010. Between October 2010 and June 2013 she worked at Plumpton again and lately managed research collaborations with Jon Danielewicz and Professor Richard Marchal and established new research links with the University of Brighton's faculty of Science and Engineering.

In July 2013, the Cool Climate Oenology and Viticulture Institute (CCOVI) at Brock University in St. Catharines, Ontario, Canada, recruited Belinda Kemp in order to apply her extensive research and outreach experience to the Canadian grape and wine industry.[1]

10.2 References

[1] Oenologist arrives to address grape and wine research priorities

Chapter 11

Mia Klein

Mia Klein is an Californian winemaker. Before studying enology at the University of California, Davis, Klein was a professional chef in San Francisco. Mentored by well-known Napa Valley winemakers, Cathy Corison and Tony Soter, Klein has made wine for several critically acclaimed wineries including Chappellet Winery, Araujo Estate Wines, Cimarossa, Spottswoode, Viader, Dalla Valle Vineyard and Robert Pepi Winery.

She is currently producing wine under her own-labeled, Selene, as well as producing wine for Bressler Winery and Boyanci.

11.1 Personal life and education

Mia Klein grew up in Hermosa Beach, California.[1] When she was a senior in high school her family relocated to San Francisco. She worked at a wine shop in San Francisco.[2] Klein was a chef working in a seafood restaurant.[3] While at work, she would taste wine, which invoked her interest in wine. She started studying enology at the University of California, Davis.[4] She graduated in 1983.[2]

11.2 Career

After graduating with her degree in Enology, she worked at Chappellet Winery under winemaker Cathy Corison.[4] In the early 1980s she became assistant winemaker.[2] After Chappellet, she worked at Robert Pepi Winery. She met Tony Soter, while working at Robert Pepi. Soter and Klein would go start a consulting business, providing services numerous Napa Valley wineries. Klein also worked at Dalla Valle Vineyard. She replaced winemaker Heidi Peterson Barrett at Dalla Valle.[4] Klein was the first winemaker for Bressler Winery in 1999. She is the winemaker for Boyanci and Cimarossa Vineyards, and has provided consulting services to Palmaz Vineyards, Spottswoode, Viader and cult wine label Araujo.[2][3][5] Soter left the consulting business in 1999 and Klein continues to consult.[6]

11.2.1 Selene Wines

Klein started her own wine label, Selene Wines, in 1991. The label makes Merlot, Sauvignon blanc, Cabernet Sauvignon, and a red wine blend.[4] The Merlots are macerated in tanks with skins being kept on for 21-days. The wine is aged in 70% new French Oak barrels. The Napa Valley Merlot's grapes are sourced from grapes in St. Helena and near Calistoga.[3] The Sauvignon blanc, Selene Sauvignon blanc, uses Musqué grapes from the Hyde Vineyard in the Carneros AVA.[3][4] The first vintage was in 1992.[7] The wine is not processed via malolactic fermentation. It's fermented, in 40% to 50% new French oak barrels, and half in stainless steel barrels.[3] The blend is called Chesler. The wines are made at the Laird Family Estate winery.[4] A large portion of the Merlot and Sauvignon blanc wines are distributed to restaurants.[3]

11.3 References

[1] Neff, Kristen Jones. "Wonderful Women Winemakers". Edible. Retrieved 9 December 2012.

[2] Teague, Lettie. "Napa's Hired Gun". Food & Wine. Retrieved 9 December 2012.

[3] "Serious Merlot". Wine People. Retrieved 9 December 2012.

[4] "Mia Klein". *People and institutions*. Calwineries. Retrieved 9 December 2012.

[5] Gilbert, Lucia and John. "Mia Klein". *Winemakers A-Z*. Women Winemakers of California. Retrieved 9 December 2012.

[6] Woolever, Laurie. "On Harvest with ... Mia Klein". *Seasonal*. Wine Spectator. Retrieved 9 December 2012.

[7] "Salene Wines". *North Bay Biz*. June 2006. Retrieved 20 December 2012.

11.4 External links

- Mia Klein on Twitter

- Official website

Chapter 12

Max Léglise

Max Léglise (1924–1996) was a French oenologist. Most of his career was spent working at the *Station œnologique de Bourgogne* (INRA, Beaune), where he entered in 1948, and that he directed from 1962 to 1984.

Departing from the dominant conventional œnology that he practiced at the beginning of his career, he developed biological methods to be applied to vinification. He was also one of the initiators of sensory analysis and was therefore well regarded by fellow oenologists, wine merchants, and restaurateurs.

12.1 Selected publications

- *Une initiation à la dégustation des grands vins* (1976)

- *Les méthodes biologiques appliquées à la vinification & à l'oenologie* (2 volumes)

- *La vigne et le vin entre Ciel et Terre* (text of a conference held in 1990)

12.2 See also

- List of wine personalities

Chapter 13

Zelma Long

Zelma R. Long (born c. 1945) is an American enologist and vintner. She is considered to be one of the female pioneers of wine production in the U.S. state of California, and was the first woman to assume senior management of a Californian winery, Simi Winery, of which she was president from 1989 to 1996.[1] Long founded and was the first president of the American Vineyard Foundation to help finance research in enology and viticulture and also founded the American Viticulture and Enology Research Network (AVERN). She is the co-owner of Long Vineyards in St. Helena, California, and the Vilafonte Wine Estate in South Africa. Long has particularly been active in research into viticulture in Washington state.

13.1   Career

Long graduated from the Oregon State University in 1965 and had an internship at the University of California San Francisco Medical Center, followed by a brief career as a dietician.[2][3] In 1966, her family purchased a hillside vineyard in the Napa Valley and her love of viticulture was born.[2] She then pursued a master's degree in Enology and Viticulture at the University of California, Davis in 1968 but left there in 1970 before completing her degree to work a harvest at Robert Mondavi Winery in the Napa Valley, ultimately being promoted to chief enologist during the boom of the 1970s, between 1973 and 1979.[3] Robert Mondavi considered her departure from the company in 1979 as one of his biggest losses.[4] She became Vice President of Simi Winery in the Alexander Valley in 1979,[5] and after enrolling at Stanford University, became President and CEO from 1989 to 1996.[1][3] She is credited with breathing new life into Simi and modernizing it, introducing new winemaking techniques and establishing new vineyards.[6][1] She then served as Executive Vice President at Chandon Estates until 1999 when she committed herself to running Long Vineyards, which she established with her then-husband Bob Long in St. Helena, California in 1977.[3][2]

Long founded and was the first president of the American Vineyard Foundation (AVF) to help finance research in enology and viticulture, and also founded the American Viticulture and Enology Research Network (AVERN).[1][7] Long has been dedicated to enology research into other states, particularly Washington which she considers to have the best Merlot grapes in the world.[3] She established Zelphi Wines with current husband Philip Freese, and has since ventured into new wine projects in Germany and South Africa, establishing the Vilafonte Wine Estate in Paarl, South Africa.[3] She has said of winemaking, "Age is a precondition to greatness. Even if only 10 percent of all fine wines are fully mature before they are drunk up, it's those great aged wines that set the standard."[8]

As one of the female pioneers of the development of viticulture in the United States, she is highly acclaimed, and has been a mentor to numerous aspiring female vintners in California.[9][1][10] In a male-dominated industry, her coworkers have named her "Miss Oblivious", because of her "ability to turn a deaf ear to sexism".[11] The Amis du vin Association cited her as a "superior" winemaker.[12] She was a recipient of the James Beard Award for Wine Professional of the Year in 1997, California Wine pioneer by *Wine Spectator* in 1993, the MASI award for her contribution to international wine in 1994, and that year she was also nominated for Woman of the Year.[7][3] In January 2009, the University of California, Davis honored her with The Outstanding Alumni award.[7] She is the author of books such as *Enological and Technological Developments*.[13] She has a keen interest in Buddhist, Asian and African art and culture, and enjoys birdwatching and horseriding in her spare time.[7]

13.2   References

[1] Matasar, Ann B. (2006). *Women of Wine: The Rise of Women in the Global Wine Industry*. University of California Press. p. 90-91. ISBN 978-0-520-93070-4. Retrieved 19 March 2013.

[2] "Who's Who". Long Vineyards. Retrieved 19 March 2013.

[3] "Zelma Long". Wine Behind the Label. Retrieved 19 March 2013.

[4] Simon, Joanna (24 September 1990). *Harrods book of fine wine*. Mitchell Beazley. p. 194. ISBN 978-0-85533-788-9. Retrieved 19 March 2013.

[5] *Playbill*. American Theatre Press. 1988. p. 79. Retrieved 19 March 2013.

[6] *Wine & Spirits*. Winestate Publications. 2007. p. 118. Retrieved 19 March 2013.

[7] "Biography of Zelma Long" (PDF). Asev.org. Retrieved 19 March 2013.

[8] New York Media, LLC (19 February 1990). *New York Magazine*. New York Media, LLC. p. 69. Retrieved 19 March 2013.

[9] Haeger, John Winthrop (17 November 2008). *Pacific Pinot Noir: A Comprehensive Winery Guide for Consumers and Connoisseurs*. University of California Press. p. 3. ISBN 978-0-520-94211-0. Retrieved 19 March 2013.

[10] Ausmus, William A. (31 May 2008). *Wines and Wineries of California's Central Coast: A Complete Guide from Monterey to Santa Barbara*. University of California Press. p. 270. ISBN 978-0-520-93183-1. Retrieved 19 March 2013.

[11] Monaghan, Patricia (14 October 2008). *Wineries of Wisconsin and Minnesota*. Minnesota Historical Society. p. 150. ISBN 978-0-87351-708-9. Retrieved 19 March 2013.

[12] *Wine*. Les Amis du Vin. 1987. p. 76. Retrieved 19 March 2013.

[13] Sullivan, Charles Lewis (1 October 1998). *A Companion to California Wine: An Encyclopedia of Wine and Winemaking from the Mission Period to the Present*. University of California Press. p. 283. ISBN 978-0-520-21351-7. Retrieved 19 March 2013.

Chapter 14

Justin Meyer

Justin Meyer (born **Raymond Meyer**, 11 November 1938 – 6 August 2002) was an American vintner, enologist, and monk of the Christian Brothers. He was the founder along with Ray Duncan of Silver Oak Cellars in 1972, a successful winery based in the Napa Valley and Alexander Valley. Today Duncan's sons David Duncan and Tim Duncan run Silver Oak Cellars, as well as Twomey Cellars, established in 1999. Meyer sold his share of the company to Duncan in 2001.[1][2] One of California's top wine experts, he was President of the American Vineyard Foundation in the 1990s and also held numerous other positions in the wine industry. The *San Francisco Chronicle* cites Meyer as "one of the legends of the Napa Valley".[3]

14.1 Background

Meyer was born Raymond Meyer on 11 November 1938 in Bakersfield, California.[4] Just out of high school, he became a monk of the Christian Brothers, and changed his first name to Justin. He taught Spanish at a Christian Brothers high school in Sacramento and in 1964 was apprenticed to winemaker Brother Timothy at their winery, Greystone Cellars, in St. Helena, California in the Napa Valley.[5][4] At one point the Christian Brothers ran 6 wineries and were the largest brandy producers in the world according to Meyer.[4]

14.1.1 Career

He left the Christian Brothers in 1972, and became president of V&E Consulting and Management Company.[6] That year, Meyer met Colorado entrepreneur Ray Duncan who had purchased a 750 acre plot of land in the Napa Valley of northern California, which was formerly the Oakville Dairy farm, as an investment in growing and selling grapes.[7][8] Meyer, a winemaker who was a monk of the Christian Brothers religious order, formed an agreement, with Duncan, setting up a winery on the Christian Brother's site in St. Helena. Meyer would provide his winemaking,

Meyer with Silver Oak Cellars co-founder, Ray Duncan

cultivation and Californian market expertise while Duncan provided the financial backing.[8]

The pair bottled their first vintage Cabernet Sauvignon in 1972, aging the wine in the old Keig Dairy barn on the original plot of land Duncan purchased.[9] They made the decision to produce only Cabernet Sauvignon and to attempt to produce the finest in the world,[10][11] to age the wine exclusively in American oak barrels.[12] Meyer said of the reason for deciding to concentrate on one wine, "it was kind of a reaction to my days at Christian Brothers, where we made so many wines it was hard to do them all right, and it was kind of in keeping with what I thought — that Cabernets were what Napa and Sonoma did best, so why not devote our attention to that. This is a pretty common concept in France."[2] According to noted wine critic Robert Parker, Meyer always believed in harvesting ripe, physiologically mature fruit.[13] In answer to the question of why Silver Oak insists upon aging their wines in American oak barrels, Meyer once said, "I'm happy with a cellar of about 65 degrees. Aging is speeded up by heat and slowed down by cold, but the only thing I do to modify my cellar is drink it faster... To my palate, American oak imparts less wood tannin than French oak. I like tannic wine about as much as I like tough steak."[14]

Meyer and Silver Oak's barrels

Meyer and Duncan made their first three vintages at the Christian Brothers winery, and in 1975 bought the Franciscan Winery, selling it in 1978 to buy the Silver Oak winery near Oakville and buying up new land.[8] The Silver Oak winery began production in 1981 and the 1982 harvest was considered by Meyer to be "something special" and was attempted to be replicated in later years.[8] Growth of the company from 1977 onwards enabled Silver Oak to purchase further vineyards in the 1980s and early 1990s, becoming one of the most successful Cabernet Sauvignon brands.[15][16] By the 1994 vintage, the concept changed and the Silver Oak Napa became a blend of Cabernet Sauvignon, Cabernet Franc, Merlot and Petit Verdot, but still aged solely in American oak, a California take on the clas-

sic Bordeaux chateau-bottled red wine.[17] Meyer was President of the American Vineyard Foundation in the 1990s, and also held several other important positions in the wine industry.[18] He breathed new life into the flailing American Vineyard Foundation from the late 1980s; *Wines and Vines* stated that Meyer's most important contribution will be his influence in bringing growers and winemakers together to shape the very future of California wine." He trained Daniel Baron extensively to replace himself as Silver Oak's chief winemaker. In the late 1990s, he was diagnosed with Type-2 Diabetes and a degenerative brain disease.[3][19] In January 2001 he sold his share of the company to Ray Duncan, citing health problems, [5] but continued as a consulting winemaker until his death in August 2002.[20][21] Meyer once said "Only one wine can be your best, and I felt that cabernet was what we did best in Napa and Sonoma".

14.2   Death and legacy

Meyer enjoying a glass of wine

Meyer died of a heart attack at the age of 63 while on vacation in the Sierra Nevada mountains near Lake Tahoe on

6 August 2002.[22][4] A memorial service was held at Silver Oak on 10 August.[23]

Upon his death, Jim Wolpert, the head of the department of viticulture and enology at University of California Davis said "The debt of gratitude that we in the research community owe to him is immeasurable. He will be missed in ways that I don't think we all understand yet. There was no question about his reputation. His approach was always a no-nonsense, no-politics approach. He never let the main issues get sidetracked."[21] John De Luca, President of the Wine Institute said "I'm not sure people truly perceive his extraordinary impact on the wine industry. He is one of the defining figures, one of the great figures in wine of our time."[21] Patrick Gleeson, executive director of the American Vineyard Foundation, who Meyer had mentored as a young lad said "His vision and love for wine has undoubtedly made him a living legacy amongst friends, peers and those in the industry. Justin's passing will not diminish the influence he has had on the wine industry and hopefully it will inspire others to follow in his footsteps."[21]

He is survived by his wife, Bonny. The couple had two sons, Chad and Matt, and a daughter, Holly.[4] Bonny Meyer operates Bonny's Vineyard,[24][25][3] and his son Matt also operates Meyer Family Cellars,[26] which his father helped him establish in the Yorkville Highlands of Mendocino County in January 1999.

14.3 References

[1] "Colorado Business Hall of Fame: Ray Duncan". *ColoradoBiz*. 1 January 2012. Retrieved 19 March 2013.

[2] "The cult of cabernet". *ColoradoBiz*, accessed via HighBeam Research. 22 May 2009. Retrieved 19 March 2013.

[3] "SON OF SILVER OAK / The scion of one of Napa's biggest legends makes a name for himself in Mendocino County with Meyer Family Cellars". *San Francisco Chronicle*. 6 April 2007. Retrieved 19 March 2013.

[4] "Justin Meyer, 63, Winemaker Renowned for His Cabernet". *The New York Times*. Retrieved 26 March 2013.

[5] "Justin Meyer, 63; Founder of Napa Valley's Silver Oak Cellars". *Los Angeles Times*. 10 August 2002. Retrieved 26 March 2013.

[6] *Australian Viticulture*. Winetitles : printed by Hyde Park Press. 2003. p. 73. Retrieved 26 March 2013.

[7] Moss, Austin Peter (March 1991). *Cole's insiders guide to the wines & vines of Napa County*. Cole Pub. Co. p. 222. ISBN 978-0-929635-06-4. Retrieved 26 March 2013.

[8] "The cult of Cabernet: Silver Oak Cellars soars with single-minded niche.(Silver Oak Wine Cellars)(Company overview)". *ColoradoBiz*, accessed via HighBeam Research. 1 September 2007. Retrieved 26 March 2013.

[9] Kushman, Rick; Beal, Hank (16 April 2007). *A Moveable Thirst: Tales and Tastes from a Season in Napa Wine Country*. John Wiley & Sons. p. 247. ISBN 978-0-471-79386-1. Retrieved 26 March 2013.

[10] Martin, Don W.; Martin, Betty Woo; Sebastiani, Sam (2 March 2005). *The Best of the Wine Country*. DiscoverGuides. p. 138. ISBN 978-0-942053-43-2. Retrieved 26 March 2013.

[11] Lenkert, Erika (19 July 2006). *Frommer's Portable California Wine Country*. John Wiley & Sons. p. 59. ISBN 978-0-470-04313-4. Retrieved 26 March 2013.

[12] Jennings, Richard (6 June 2012). "Silver Oak: Reappraising a California Cabernet Icon". *Huffington Post*. Retrieved 26 March 2013.

[13] Parker, Robert (1996). *The Wine Buyer's Guide*. Dorling Kindersley. p. 978. ISBN 978-0-7513-0342-1. Retrieved 26 March 2013.

[14] "Drinking in the sights, wines at Yosemite". *Daily Herald*, accessed via HighBeam Research. 13 January 1999. Retrieved 26 March 2013.

[15] Dorling, Kindersley; Gordon, Jim (1 October 2010). *The Wine Opus*. Dorling Kindersley Limited. p. 37. ISBN 978-1-4053-5267-3. Retrieved 26 March 2013.

[16] *Quarterly Review of Wines*. Richard L. Elia. 2005. p. 54. Retrieved 27 February 2013.

[17] Boyd, Gerald D. (27 May 2009). "The "New" Silver Oak Napa". Wine Review Online. Retrieved 26 March 2013.

[18] Perdue, Lewis (1 June 1999). *The Wrath of Grapes: The Coming Wine Industry Shakeout And How To Take Advantage Of It*. Lewis Perdue. p. 54. ISBN 978-0-380-80151-0. Retrieved 26 March 2013.

[19] "Justin Meyer". Telegraph-Herald (Dubuque), accessed via HighBeam Research. 13 August 2002. Retrieved 26 March 2013.

[20] "Justin Meyer Sells Share Of Silver Oak Wine Cellars". *Wines & Vines*, accessed via HighBeam Research. 1 February 2001. Retrieved 26 March 2013.

[21] "Justin Meyer's legacy to California wine". *Wines & Vines*, accessed via HighBeam Research. 1 November 2002. Retrieved 26 March 2013.

[22] *Vineyard & Winery Management*. Vineyard & Winery Management. 2002. p. 154. Retrieved 26 March 2013.

[23] "Justin Meyer dead at 63". *Wines & Vines*, accessed via HighBeam Research. 1 September 2002. Retrieved 26 March 2013.

[24] Clarke, Oz (1 October 1991). *Oz Clarke's new classic wines.* Simon & Schuster. p. 125. ISBN 978-0-671-69620-7. Retrieved 26 March 2013.

[25] "Bonny's Vineyard". Bonny's Vineyard. Retrieved 26 March 2013.

[26] "Family tree is a grapevine" (PDF). Mfcellars.com. Retrieved 26 March 2013.

Chapter 15

Hermann Müller (Thurgau)

For other Hermann Müllers: see Hermann Müller (disambiguation).

Hermann Müller birthplace, in Tägerwilen/Thurgau

15.1 Biography

Hermann Müller was born to Konrad Müller, a master baker and vintner, and his wife Maria Egloff, the daughter of Karl Anton Egloff, a wine merchant of Oestrich, Hessen. He attended the Lehrerseminar Kreuzlingen (Kreuzlingen Teachers College) (1869-70). He taught in Stein am Rhein (1870-72) while studying at the Polytechnikum Zürich (1872 graduate). He then attended the University of Würzburg for graduate studies under Julius von Sachs, was awarded his PdD in 1874 and stayed some time as Sachs' assistant. During the years 1876–1890 he worked at the Prussian Institute for Horticulture and Viticulture (*Königlich Preussische Lehranstalt für Obst- und Weinbau*) in Geisenheim, Rheingau where he led its experimental station for plant physiology.

In 1891 he returned to Switzerland as director of the newly created Experimental Station and School for Horticulture and Viticulture (*Versuchsstation und Schule für Obst-, Wein- und Gartenbau*) in Wädenswil, where he stayed until his 1924 retirement. From 1902, he was also connected to Polytechnikum Zürich as professor of botany.

He worked on teams which investigated fertility of the vine, vine diseases, and malolactic fermentation in wine.

Hermann Müller (Thurgau)

Hermann Müller (21 October 1850 in Tägerwilen, Thurgau, Switzerland - 18 January 1927, in Wädenswil, Zurich), was a Swiss botanist, plant physiologist, oenologist and grape breeder.[1][2] He called himself Müller-Thurgau, taking the name of his home canton.

In 1890, he was made an honorary member of the German Viticultural Association and in 1920 he received an honorary doctorate from the University of Bern.

Müller researched and published on a wide range of topics in viticulture and winemaking, including the biology of vine flowering, assimilation of nutrients by the vine, vine diseases, alcoholic fermentation of wine, breeding of strains of yeast with specific properties, malolactic fermentation, development of wine faults, and methods for producing alcohol-free grape juice.

15.2 Breeding of the Müller-Thurgau grape variety

Main article: Müller-Thurgau

During his time in Geisenheim, Müller created the grape variety Müller-Thurgau in a breeding programme initiated in 1882, by crossing Riesling with Madeleine Royale, although for a long time it was erroneously assumed to be Riesling x Silvaner. Müller's goal was to combine the aromatic properties of Riesling with the earlier and more reliable ripening of Silvaner. Experimental plantations continued in Geisenheim until 1890, and in 1891 150 plants were shipped to Wädenswil where trials continued under Heinrich Schellenberg (1868–1967). The most successful clone of the trials (serial no. 58) was propagated in 1897 under the designation Riesling x Silvaner 1. Vines of this variety were distributed in Switzerland and abroad from 1908, and in 1913, 100 vines of this variety were taken to Germany by August Dern (1858–1930), who had worked with Müller in Geisenheim. Dern introduced the name "Müller-Thurgau" for the variety, while Müller himself continued to call it Riesling x Silvaner 1, although he did express doubts that this was the actual parentage of the new variety, and speculated that some misidentification of vine material could have occurred in the move from Geisenheim to Wädenswil.[3][4]

Many experimental plantations of Müller-Thurgau in Germany were conducted from 1920, and its breakthrough from 1938 is credited to the grape breeder Georg Scheu in Alzey.[4] By the 1950s it had become the most cultivated of any newly created grape varieties. It was the most planted grape variety of Germany from the late 1960s to the mid-1990s, and is still the second-most planted.

15.3 External links

15.4 References

[1] Historisches Lexikon der Schweiz: Müller [Müller-Thurgau], Hermann (German)

[2] Gesellschaft für Geschichte des Weines: Müller-Thurgau, Hermann (1850-1927) (German)

[3] Wein-Plus Glossar: Müller-Thurgau, accessed 23 January 2013

[4] 125 Jahre Müller-Thurgau, accessed 14 October 2009 (German)

[5] "Author Query for 'Müll.-Thurg.'". International Plant Names Index.

Chapter 16

List of oenologists

This list is incomplete; you can help by expanding it.

This is a **list of notable oenologists**.

16.1   Oenologists

- Alberto Antonini
- Miguel Brascó
- Cathy Corison
- Tullio De Rosa
- Peter Gago
- Hermann Jaeger
- Max Léglise
- Zelma Long
- Justin Meyer
- Hermann Müller (Thurgau)
- Ottavio Ottavi
- Jacques Puisais
- Michel Rolland
- Carol Shelton
- Păstorel Teodoreanu
- Miguel A. Torres
- Keith Wallace

16.2   See also

- Lists of people by occupation

Chapter 17

Ottavio Ottavi

Ottavio Ottavi.

Ottavio Ottavi (15 August 1849 – 12 January 1893) was an Italian oenologist.

17.1 Biography

Ottavi was born in Sandigliano. His father Giuseppe Antonio Ottavio was an agronomist,[1] and his brother Edoardo, editor of the journal *Il Coltivatore*, was also seen as a significant figure in the development of nineteenth-century Italian viticulture.[2]

He was the author of various treatises and monographs, including *Enologia teorico-pratica* (1898), and was the founder of the *Giornale vinicolo italiano*. His *Inno ai Krumiri* (1886) is a "hymn" to the krumiro, a type of biscuit created in Casale Monferrato, the town where he largely lived and where he died in 1893.

17.2 References

[1] Félix Sahut, *Les vignes américaines: leur greffage et leur taille* (Monpelliers: Camille Coulet; Paris: A Delahaye et Lecrosnier, 1887), p. 503.

[2] Coulet, pp. 509 and 714.

17.3 External links

- Photograph and details of his commemorative plaque, sculpted by Leonardo Bistolfi, in Casale Monferrato.

Chapter 18

Émile Peynaud

Émile Peynaud (June 29, 1912 – July 18, 2004) was a French oenologist and researcher who has been credited with revolutionizing winemaking in the latter half of the 20th century, and has been called "the forefather of modern oenology".[1]

18.1 Biography

Peynaud entered the wine trade at the age of fifteen with the négociant Maison Calvet.[1] At Calvet he worked under the chemical engineer Jean Ribéreau-Gayon, and they developed methods of analysing the wines that were to be purchased. In 1946, Peynaud completed his Doctorate at the University of Bordeaux and joined its faculty as a lecturer. Ribéreau-Gayon at this time was also teaching at the University, and the two shifted their previous focus of problems faced by Calvet to the problems faced by the winemakers themselves.

While at the University of Bordeaux, where he became a professor of oenology, Peynaud worked at providing scientific explanations for many problems encountered in the process of winemaking. He convinced the wineries to begin picking of grapes at vineyards up to two weeks later than usual, and to complete the picking as quickly as possible. The practice of also picking underripe or rotten grapes was abandoned, so that the selected fruit arriving at the winery was of the best possible quality.[2]

Peynaud introduced crushing and fermenting fruit in separate batches based on vine age, vineyard location, or any other factor that resulted in fruit of differing qualities in order to control tannin extraction. He then applied the cool fermentations used in Champagne to still white Bordeaux in order to control fermentation temperatures.

Proposing methods that ran counter to many traditions, in the 1950s and 1960s skeptics would use the term "Peynaudization" of Bordeaux, but as his advice usually produced superior wines, criticism came to an end.[2]

Peynaud considered the control of malolactic fermentation to be one of his most important contributions to winemaking. It was commonly believed that malolactic fermentation was a sickness. He helped the wineries realize that they needed to encourage and control malolactic fermentation. He also stated, "Using only the very best grapes is a new phenomenon," considering this "the crowning achievement of [his] work."[2]

Peynaud was the *Decanter* Man of the Year in 1990.[1] Michel Rolland is one of his pupils.

18.2 Selected bibliography

- Peynaud, Émile; J. Blouin (2005) [1971]. *Connaissance Et Travail Du Vin* (in French) (4th ed.). Dunod. ISBN 2-10-049296-9.

- Peynaud, Émile (1984). *Knowing and Making Wine.* trans. Alan Spenser. Wiley-Interscience. ISBN 0-471-11376-X.

- Peynaud, Émile; J. Blouin (1996) [1983]. *Le goût du vin* (in French). Paris: Dunod. ISBN 2-10-002750-6.

- Peynaud, Émile; intro. by Michael Broadbent, M.W. (1996). *The Taste of Wine: The Art and Science of Wine Appreciation.* trans. Michael Schuster. London: Macdonald Orbis. ISBN 0-471-11376-X.

18.3 See also

- List of wine personalities

18.4 References

[1] Styles, Oliver, *Decanter* (2004-07-20). "Emile Peynaud dies at 92".

[2] Steinberger, Mike, *Slate* (2004-07-30).   "The distorted
    legacy of Émile Peynaud".

Chapter 19

Jacques Puisais

Jacques Puisais (born in Poitiers en 1927) is a French oenologist and taste philosopher. Holder of a PhD in chemistry, he directed the *laboratoire départemental et régional d'analyse* in Tours He starts giving courses of taste education in 1964. He's a member of the INAO. He created the Institut Français du Goût in 1976, in order to develop multidisciplinary research around taste and food sensitivity. Eager to introduce children to taste, he developed a method of sensory awakening, that has been used in classrooms since then.[1]

19.1   See also

- List of wine personalities

19.2   References

[1] "Classes du goût".

Chapter 20

Michel Rolland

Michel Rolland (born December 24, 1947 in Libourne, France) is an influential Bordeaux-based oenologist, with hundreds of clients across 13 countries and influencing wine style around the world. "It is his consultancies outside France that have set him apart from all but a handful of his countrymen." It is frequently addressed that his signature style, which he helps wineries achieve, is fruit-heavy and oak-influenced, a preference shared by influential critic Robert Parker.

Rolland owns several properties in Bordeaux, including Château Le Bon Pasteur, Château Bertineau Saint-Vincent in Lalande de Pomerol, Château Rolland-Maillet in Saint-Émilion, Château Fontenil in Fronsac, and Château La Grande Clotte in Lussac-Saint-Émilion as well as joint venture partnerships with Bonne Nouvelle in South Africa, Val de Flores in Argentina, Rolland Galarreta in Spain and Yacochuya (Salta) and Clos de los Siete in Argentina.

20.1 Education and early career

Vineyards at the Rolland family estate of Château Le Bon Pasteur in Pomerol.

Born into a wine making family, Rolland grew up on the family's estate Château Le Bon Pasteur in Pomerol.[1] After high school, Rolland enrolled at Tour Blanche Viticultural and Oenology school in Bordeaux with his father's encouragement. Excelling in his studies, he was one of five student chosen by director Jean-Pierre Navarre to evaluate the program's quality against that of the prestigious Bordeaux Oenology Institute. Rolland later enrolled in the Institute, where he met his wife and fellow oenologist, Dany Rolland, and graduated as part of the class of 1972.

At the Institute, Michel Rolland studied under the tutelage of renowned oenologists Pierre Sudraud, Pascal Ribéreau-Gayon, Jean Ribéreau-Gayon, and Émile Peynaud. Rolland has said these men were a great influence upon him and considers them the "Fathers of Modern Oenology."

In 1973, Rolland and his wife bought into an oenology lab on the Right Bank of Bordeaux in the town of Libourne. They took over full control of the lab in 1976 and expanded it to include tasting rooms. By 2006 the Rolland's lab employed 8 full-time technicians, analyzing samples from nearly 800 wine estates in France each year.[2] Rolland's two daughters, Stéphanie & Marie, also work at the lab.

Michel Rolland's first clients included the Bordeaux Châteaux Troplong Mondot, Angélus, and Beau-Séjour Bécot. An early setback was the loss of two Saint-Émilion first growths, Château Canon and Château La Gaffelière, due to conflict in style with the owners and Rolland. According to Rolland, the loss "calmed him down" and brought him out of an awkward stage in his early career. Twenty years later, the two chateaux returned to be part of the more than 100 wineries who employ Michel Rolland as their consultant.

In his book *Noble Rot: A Bordeaux Wine Revolution*, William Echikson writes that before Michel Rolland became consultant to Château Lascombes, it "produced about 500,000 bottles of mediocre wine, about half of which was sold not as Lascombes itself, but as the inferior Chevalier de Lascombes." Today, Echikson contends, that even the Chevalier (the second wine of the estate) is better than the old full-fledged Lascombes.

20.2 Media exposure

Michel Rolland's signature on a bottle of the Argentine wine Yacochuya which Rolland owns as part of a joint venture in Cafayate.

Rolland features prominently in the critical 2004 documentary *Mondovino* by Jonathan Nossiter as an agent of wine globalization. In *Mondovino*, Rolland is seen on several occasions advising his clients to microoxygenate their wines, including a scene at Château Le Gay in Bordeaux. Since the film, Rolland has said that he is "not a fan of microoxygenation. The film suggests I am. Some of my clients inquire about it. It can help in special conditions — if the tannins are fierce or hard, micro-oxygenation can make them softer and rounder. In certain countries with certain terroir, like Chile or Argentina, I may use it." James Suckling, formerly of *Wine Spectator*, notes in an article about Rolland that "He is not a proponent of micro-oxidation in wine-making as some suggest, and never has been".[2]

Michel Rolland is also a wine making consultant for the Amphorae Winery in Israel (marketed as Makura in the United States) and has started signing his name to their premium Makura series. He visits Amphorae and their vineyards once a year and has his assistants throughout the year help implement his practices adopted by Amphorae's wine making team at the winery and in their vineyards.

Rolland is among the wine personalities satirised next to

Robert Parker in the 2010 *bande dessinée* comic book, *Robert Parker: Les Sept Pêchés capiteux*.[3][4]

20.3 Influence

From his consulting work and media presence, Michel Rolland has influenced many aspects of both the French and global wine industry. Among the prominent wine personalities that have been influenced by Rolland is the Rhone wine producer Jean-Luc Colombo.[5]

20.4 Bordeaux vineyards under Rolland influence

Rolland holds decisive roles (such as owner, cellar master, oenologist, consultant) in a number of vineyards in Bordeaux. These include : Angélus, St-Emilion GC; Armens, St-Emilion GC; Ausone, St-Emilion GC; Beauregard, Pomerol; Bellefont-Belcier, St-Emilion GC; Bellevue Mondotte, St-Emilion GC; Blason de l'Evangile, Pomerol; le Bon Pasteur, Pomerol; Bonalgue, Pomerol; Branas Grand Poujeaux, Moulis; Brillette, Moulis; de Camensac, Haut-Médoc; Cap de Faugères, Côtes de Castillon; Certan de May de Certan, Pomerol; Chapelle d'Ausone, St-Emilion GC; Clarke, Listrac-Médoc; la Clémence, Pomerol; Clément-Pichon, Haut-Médoc; Clinet, Pomerol; Clos des Jacobins, St-Emilion GC; Clos du Clocher, Pomerol; Clos l'Eglise, Pomerol; Clos les Lunelles, Côtes de Castillon; Clos Saint-Martin, St-Emilion GC; la Commanderie de Mazeyres, Pomerol; Corbin, St-Emilion GC; Côte de Baleau, St-Emilion GC; la Couspaude, St-Emilion GC; le Crock, St-Estèphe; Croix de Labrie, St-Emilion GC; Destieux, St-Emilion GC; Destieux, St-Emilion GC; la Dominique, St-Emilion GC; l'Evangile, Pomerol; Faugères, St-Emilion GC; Faugères Cuvée Péby, St-Emilion GC; la Fleur de Boüard, Lalande de Pomerol; la Fleur de Gay, Pomerol; Fombrauge, St-Emilion GC; Fontenil, Fronsac; Franc-Mayne, St-Emilion GC; la Garde, Pessac-Léognan; le Gay, Pomerol; Giscours, Margaux; Grand Mayne, St-Emilion GC; Grand Ormeau, Lalande de Pomerol; Grand-Pontet, St-Emilion GC; les Grandes Murailles, St-Emilion GC; les Grands Chênes, Médoc; la Gravière, Lalande de Pomerol; Jean de Gué, Lalande de Pomerol; Kirwan, Margaux; Larmande, St-Emilion GC; Larrivet-Haut-Brion, Pessac-Léognan; Lascombes, Margaux; Latour-Martillac, Pessac-Léognan; Léoville-Poyferré, St-Julien; Loudenne, Médoc; Magrez-Fombrauge, St-Emilion GC; Malartic-Lagravière, Pessac-Léognan; Malescot-Saint-Exupéry, Margaux; Monbousquet, St-Emilion GC; Pape Clément, Pessac-Léognan; Pavie, St-Emilion GC; Péby Faugères, St-Emilion GC; Petit Village, Pomerol; Phélan-Ségur, St-Estèphe; le Plus de la

Fleur de Boüard, Lalande de Pomerol; Pontet-Canet, Pauil-
lac; Ripeau, St-Emilion GC; Rochebelle, St-Emilion GC;
Rouget, Pomerol; la Sérénité, Pessac-Léognan; Smith Haut
Lafitte, Pessac-Léognan; la Tour-Carnet, Haut-Médoc;
Troplong-Mondot, St-Emilion GC; de Valandraud, St-
Emilion GC; la Violette, Pomerol; Virginie de Valandraud,
St-Emilion GC;

20.5   See also

- Parkerization of wine

20.6   Sources

- Echikson, William. *Noble Rot: A Bordeaux Wine Revolution.* NY: W.W.Norton, 2004.

- Robinson, Jancis (Editor) *The Oxford Companion to
  Wine.* Oxford, England: Oxford University Press, second edition, 1999

Footnotes

[1] Asimov, Eric, *New York Times* (October 11, 2006). "Satan
or Savior: Setting the Grape Standard". *The New York
Times.*

[2] Suckling, James, *Wine Spectator* "Top Gun", June 30, 2006

[3] Kakaviatos, Panos, *Decanter.com* (October 12, 2010).
Robert Parker 'honoured' by merciless cartoon satire

[4] Thunevin, Jean-Luc, thunevin.blogspot.com (October 11,
2010). Hilarious duo

[5] E. Arnold *"Winemaker Talk: Jean-Luc Colombo" Wine Spectator* April 26, 2007

Chapter 21

Carol Shelton

Carol Shelton is an American winemaker and entrepreneur. She has been called the most awarded winemaker in America and was the *San Francisco Chronicle's* Winemaker of the Year in 2005.[1][2][3]

21.1 Personal life and education

Shelton was raised in Rochester, New York and San Mateo, California. When she was a child, Shelton's mother created a memory game based around smells, Shelton credits this game as helping her have a skill in chemistry and a good nose.[3] Shelton wanted to be a paleontologist when she was ten years old. In high school she decided to become a poet.[2] Shelton studied poetry at University of California, Davis but remained undeclared in her major.[1][2] When she was a Freshman she took a tour of Sebastiani Winery. She smell of the wine cellar at Sebastiani triggered her interest in winemaking and she decided to study Enology and received her degree in 1978.[1][3][4] She was one of the first women to graduate with a degree in Enology.[2] She worked on the Aroma Wheel project under Ann C. Noble and researched yeast strains and wine.[1][2][4]

21.2 Career

After graduation, she started working at Robert Mondavi Winery. She then worked in Australia for Saltrams Wines.[1] At Mondavi, and other wineries, she wasn't allowed to work in the cellars with the men winemakers.[5][2] She almost went back to being a poet. In 1980, she started doing lab work with Andre Tchelistcheff at Buena Vista Winery, whom she credited with re-instating her interest in being a winemaker.[5] The following year she worked for Sonoma Vineyards. In 1991 she became winemaker at Windsor Vineyards. It was at Windsor where she developed a strong interest in Zinfandel wine.[1] She worked at Windsor for nineteen years.[4] Shelton left Windsor after not feeling recognized and acknowledging that the "Man-

agement was pretty male dominated and not supportive," of her work. The winery was awarded the Golden Winery Award in 1996 from the California State Fair, and despite being the winemaker for the winery, she was not acknowledged for the award.[4] During her tenure at Windsor she made 45 different wines.[3] In 2000, Shelton and her husband, Mitch Mackenzie, started Shelton-Mackenzie Wine Company. The company manages a wine label and a consulting firm called Vincare. Shelton served on the board of Zinfandel Advocate Producers from 1994 to 1998.[1]

21.2.1 Carol Shelton Wines

Shelton-Mackenzie Wine Company distributes Shelton's namesake label, Carol Shelton Wines. They make Zinfandel wine from single vineyard grapes.[1] Their Zinfandel's are called Wild Thing, Karma Zin, Black Magic Late Harvest Zinfandel, and Monga Zin.[5][6] The Monga Zin grapes come from dry farmed old vine grapes from the Cucamonga Valley AVA.[6][7] The Wild Thing is made from wild yeasts and uses grapes from Mendocino County, California.[6] Karma Zin uses old vines that are over 100 years old and come from the Russian River Valley AVA. It is 85 percent Zinfandel and a remaining blend of Alicante Bouschet, Carignane and Petite Sirah.[8] They also produce Syrah, rosé, and a dessert wine.[1] Their rosé has been labeled as Rendezvous Rosé and Mendocino County Dry Rosé.[6][9] The 2007 vintage was made from 100% Carignane.[6] As of 2012, they produced Coquille Blanc, a blend of Grenache blanc, Roussanne, and Viognier.[5] The label produces 5,000 cases a year.[1] The labels on the wine show a female character that Shelton describes as her "alterego." Their tasting room, which is by appointment, and production space is located in an industrial park in Santa Rosa, California.[6]

21.2.2 Awards

She has been the recipient of numerous awards in the wine industry. Wines that Shelton has made have been awarded

33

first place at the California State Fair.[1]

- 1993, *Bon Appétit*'s Andre Tchelistcheff Winemaker of the Year[1]

- 1996, Jerry Mead's Winemaker of the Year[1]

- 1996, Golden Winery Award, California State Fair[1]

- 1999, Dan Berger's Winemaker of the Year[1]

- 2005, *San Francisco Chronicle* Winemaker of the Year[1]

21.3 References

[1] "Carol Shelton". *Wine Information*. Calwineries. Retrieved 7 December 2012.

[2] Ryan, Sandy Fertman. "Behind the Vine: Women who own wineries". *Food & Wine*. Northside San Francisco. Retrieved 8 December 2012.

[3] Gray, W. Blake (8 December 2005). "Winemakers of the Year: Carol Shelton". *News* (San Francisco Chronicle). Retrieved 8 December 2012.

[4] "Carol Shelton". Women of the Vine. Retrieved 7 December 2012.

[5] Malouf, Mary Brown. "Sipping and chatting with winemaker Carol Shelton". Salt Lake Magazine. Retrieved 7 December 2012.

[6] Knight, James. "Wine Tasting Room of the Week". *Columns*. North Bay Bohemian. Retrieved 7 December 2012.

[7] Weeks, John (November 2008). "Small wine world after all". *McClatchy-Tribune*. Retrieved 8 December 2012. (subscription required)

[8] Melnik, Peg. "Winemaker Carol Shelton finds zins best reflect earthy origins". *Living*. Press Democrat. Retrieved 8 December 2012.

[9] Melnik, Peg. "Wine of the week: Carol Shelton, 2010 Mendocino County Dry Rose". *Living*. Press Democrat. Retrieved 8 December 2012.

21.4 External links

- Official website

- Carol Shelton Wines from Touring & Tasting

- Interview with Carol Shelton from Indigo Winepress

Chapter 22

Kilien Stengel

Kilien Stengel, born in 1972 in Nevers (Nièvre), is a French gastronomic author, restaurateur, and cookbook writer. He worked at Gidleigh Park, Nikko Hotels, Georges V Hotel in Paris, and in a lot of restaurants Relais & Châteaux (Marc Meneau, Jacques Lameloise,...). He was a teacher of Gastronomy at the Académie of Paris and of Orléans-Tours.[1] Actually, Kilien Stengel work now at the European Institute for the History and Culture of Food, in the François Rabelais University. He is captain of a culinary book fair,[2] en directot of a collection book (L'harmattan éditor). Usually, he work for Ministère de l'Éducation nationale teacher competition, Meilleur Ouvrier de France award, and Masterchef France. In 2015 his PhD (Doctorat de 3e cycle) in Information science is supervised by J-J. Boutaud.

22.1 Works

22.1.1 Actuality books

- *Alimentation Bio - Manger et boire bio*, Eyrolles publishing, 2009. ISBN 2212542518

- *Gastronomie, petite philosophie du plaisir et du goût*, collection Réflexions (im)pertinentes, Bréal publishing, 2010. ISBN 9782749509945

- *Le petit dictionnaire énervé de la gastronomie*, L'opportun publishing, 2011. ISBN 9782360750481

- *Gastronomie-Gastrosophie-Gastronomisme*, L'Harmattan publishing, 2011. ISBN 9782296551978

- *Un Ministère de la gastronomie! Pourquoi pas ?*, L'Harmattan publishing, collection *Questions contemporaines*, 2011 ISBN 9782735220687

- *Manifeste du savoir-manger - Pour que nos enfants sachent se nourrir*, Praelego publishing, 2012, ISBN 9782813101839

- *Suis-je ce que je mange ?*, Le Temps qu'il fait publishing, collection Littérature, 2012 ISBN 9782868535887

- *Une cantine peut-elle être pédagogique ?* L'Harmattan publisher, collection *Questions contemporaines*, 2012, ISBN 9782296964198

- *L'Aide alimentaire : colis de vivres et repas philanthropiques - Focus sur la Gigouillette*, au profit des Restaurants du Cœur, L'Harmattan publishing, collection *travaux historiques*, 2012 ISBN 9782296967762

- *Traité de la Gastronomie : Patrimoine et Culture*, Sang de la Terre publishing, 2012, ISBN 9782869852808

- *Ca se bouffe pas, ça se déguste*, Bréal publishing, 2013

- *Traité du vin en France : Traditions et Terroir*, Sang de la Terre publishing, 2013 ISBN 9782869853102

- *Traité du fromage en France : Caséologie et authenticité*, Sang de la Terre publishing, 2014

- *Hérédités alimentaires et identité gastronomique*, L'Harmattan publishing, 2014 ISBN 9782343020969

- *Le lexique culinaire de Ferrandi : tout le vocabulaire de la cuisine et de la pâtisserie expliqué en 1.500 définitions et 200 photographies*, Éditions Hachette, 2015, ISBN 978-2011775993

- *Dictionnaire du Bien Manger et des Modèles Culinaires*, collection "les dictionnaires", Éditions Honoré Champion, 2015, ISBN 978-2745330581

22.1.2 History books

- *La Gastronomie du produit à l'assiette*, Éditions Alan Sutton, 2008.

- *Chronologie historique de la Gastronomie et de l'Alimentation* (Dictionnaire), Éditions Du Temps (diffusion Éditions du Seuil), 2008.

- *Clamecy – Événements fêtes et vie quotidienne*, collection Mémoire en images", Éditions Alan Sutton, 2010.

- *Anthologie littéraire de la gastronomie à la Belle époque* Éditions L&C, 2012

- *La gastronomie autrefois*, Sud Ouest editor, 2012 ISBN 978 2817702292

- *Histoire divertissante et curieuse de la gastronomie*, Grancher publishing, 2013 ISBN 978 2733912614

- *A table avec Jules Verne et Phileas Fogg - tour du monde en 80 recettes*, editor Agnés Vienot ISBN 9782353261512

22.1.3   Poésie books

- *Les poètes de la bonne chère*, Anthologie de poésie gastronomique, Éditions de la Table ronde, (groupe Gallimard), 2008.

- *Drôles de drames*, collectif, Codexlibris publishing, 2010.

- *Poètes du vin, Poètes divins*, préfaces de Jean-Robert Pitte (président of Paris-Sorbonne University), collection Écriture, Archipel publishing, 2012 ISBN 9782359050561

- *Permission de servir* Éditions L&C, 2012

22.1.4   Trivia books

- *Le Petit Quiz du Vin*, Éditions Dunod (groupe Hachette livre), 2007.

- *Le Grand QCM du vin*, Éditions Dunod (groupe Hachette livre), 2007.

- *Le Petit Quiz du vin – version japonaise*, éditions Sakuhin Sha (Tokyo), 2009

- *Le Grand QCM du vin – version japonaise*, éditions Sakuhin Sha (Tokyo), 2009

- *Le Grand Quiz du Fromage*, Éditions Lanore Delagrave, (Groupe Flammarion), 2008.

- *Le Grand Quiz de la bière*, Éditions Lanore Delagrave, (Groupe Flammarion), 2008. (Gourmand Cookbook Awards 2010, categorie Beerbook)

- *QG 500, le quiz de la gastronomie –Testez votre quotient culinaire*, Éditions Menu Fretin, 2009.

- *Le nouveau petit quiz du vin - 2e édition*, Éditions Dunod, 2010.

- *La Touraine en question*, co-auteur avec Patrick Prieur, Éditions Alan Sutton, 2011.

- *Pommard ou Pomerol ?*, Éditions Dunod, 2011.

- *Montmartre en question - Patrimoine et gastronomie*,co-auteur Éditions Alan Sutton, 2011

22.1.5   Games

- *La boîte à Quiz spéciale Cuisine - Testez votre quotient culinaire*, Éditions Marabout, 2011.

22.1.6   Practical books

- *Les critiques aux fourneaux*, au profit des Restos du Coeur, collectif, Éditions 4 chemins 2008 (Gourmand World Cookbook Awards2009, category Livre caritatif)

- *Œnologie et crus des vins*, Éditions Jérôme Villette (diffusion Matfer), 2008.

- *Le kit pédagogique du professeur professionnel*, éditions Eyrolles 2008.

22.1.7   School books

- *Aide-mémoire de la gastronomie en France*, Éditions BPI, 2006.

- *Technologie de service*, Éditions Bertrand Lacoste 2008.

- *Technologie culinaire*, Éditions Bertrand Lacoste 2008.

22.1.8   Part of book and direction of book

- *Étude de documents, La relation Mets-Vins*, dans *TDC n°1064 « Les Repas gastronomique des Français »*, éditions CNDP, 2013

- *L'enseignement des produits laitiers en école hôtelière : une approche plurielle*, dans *Les reconfigurations récentes des filières laitières en France et en Europe*, sous la direction de Daniel Ricard, collection CERAMAC, Presses universitaires Blaise Pascal, 2013

- *L'histoire du canard*, dans *Le Canard*, First éditions, 2014

- (Dir.) *Des fromages et des hommes Ethnographie pratique, culturelle et sociale du fromage*, l'harmattan, 2015

22.1.9 Collaborations reviews

- Directeur de la collection *Gastronomie et art culinaire*, Éditions du Temps (2006–2010)

- Responsable de publication de *Gusto*, revue culturelle, Éditions ASA (2008–2009)

- Membre comité éditorial des Presses universitaires François Rabelais

- Collaborateur à Food & History, revue scientifique européenne, Éditions Brepols

- Membre du comité de rédaction Les Cahiers de la gastronomie, revue culturelle, éditions Menu Fretin, (Prix littéraire Gastronomie-culture 2010)

22.1.10 Awards

- 2015 : The French government awarded him its highest honour, the decorations of Chevalier in Order of Agricultural Merit.

- 2014 : Prix Montesquieu, catégorie *Littérature* pour *Poètes du vin Poètes divins*.[3]

- 2014 : Nomination au prix Jean Carmet, salon du livre et du vin de Saumur, pour *Traité du vin* (Sang de la terre)

- 2013 : Gourmand World Cookbook Awards, category *Littérature gastronomique* for *Histoire divertissante et curieuse de la Gastronomie* (ed. Grancher)

- 2013 : Grand prix Académie nationale de cuisine, category *Cuisine du monde* for *A table avec Jules Verne - Le tour du monde de Phileas Fogg en 80 recettes* (ed. Vienot)

- 2012 : Gourmand World Cookbook Awards, category *Cuisine Française* for *Le Traité de la gastronomie française* (ed. Sang de la terre)

- 2012 : Mérite culinaire Prosper Montagné

- 2010 : Gourmand World Cookbook Awards for *Le Grand Quiz de la bière* (éd. Delagrave)

- 2009 : Gourmand World Cookbook Awards for *Les critiques aux fourneaux* (ed.Quatre Chemins)

- 2009 : Gourmand World Cookbook Awards for *Chronologie de la gastronomie et de l'alimentation* (ed. du temps)

- 2008 : Mention spéciale du jury du 'Salon international du livre gourmand', à Périgueux, for *Les critiques aux Fourneaux* (éd. Quatre Chemins).

22.2 References

[1] "Académie d'Orléans-Tours- Official Website". Ac-orleans-tours.fr. Retrieved 27 April 2011.

[2] {{Journal Le Monde http://blog.chefsimon.lemonde.fr/kilien-stengel-ne-en-gastronomie/}}

[3] Lauréats 2014

Chapter 23

Lane Tanner

Lane Tanner is an American winemaker and consultant. She was the second woman winemaker in Santa Barbara County when she started working in the industry in 1981.[1] She goes by the nickname "Pinot Czarina."[2]

23.1 Personal life and education

Tanner grew up in Kelseyville, California. She graduated from San Jose State University with a degree in chemistry in 1976.[3] After graduating, she started working in the air pollution industry. One winter she worked in Glendive, Montana. After Glendive, she quit her job and moved back to Kelseyville.[4] Tanner was married to the owner of The Hitching Post.[2] Tanner is now married to winemaker Ariki Hill.[5]

23.2 Career

In 1980, Tanner started working as on the bottling department, putting wine labels on bottles, at Konocti Winery in Kelseyville.[2][3][4] When the winery learned that she had studied chemistry she became a lab technician.[2] She worked there for one year and worked with Andre Tchelistcheff. She was suggested by Tchelistcheff for a job at Firestone Winery. She joined Firestone in 1981 as a winemaker.[5] At Firestone she was able to improve her skills at wine tasting and in 1984 she started her on consulting firm. Her first client was making wine for The Hitching Post.[3] She was one of the first winemakers to use Pinot grapes from Bien Nacido Vineyard.[5] In late 2010 she retired from the wine industry.[1] In 2012, Tanner was named winemaker for Sierra Madre Vineyard.[4][6] At Sierra Madre, she makes Pinot noir, Chardonnay and Pinot blanc wines.[4]

23.2.1 Lane Tanner Winery

Tanner started her own winery in 1989, Lane Tanner Winery.[3][5] The label became the first in the Central Coast region to devote itself to Pinot noir.[1] The winery is located in the Santa Maria Valley AVA and the first wine was a 1989 vintage.[3][5] The label made approximately 1,800 cases a year and specialized in Pinot noir. They also made Syrah. The Pinot noir grapes came from three vineyards in Santa Barbara County: Julia's Vineyard, Bien Nacido Vineyard, and Melville Vineyard. The Syrah grapes came from French Camp Vineyard which is located in San Luis Obispo County.[3] Tanner's wines aim to stay low in alcohol content and sulfites.[2][5] The latter is because Tanner is allergic to them.[5] The Pinot was fermented for over two weeks with yeast. It was then aged for eleven months in 46 percent French oak barrels.[5] The last vintage was in 2009.[4] The back of her Pinot bottles had the motto: "laugh more- flirt often," written on them.[7]

23.3 References

[1] Hardesty, Kathy (February 2011). "Lane Tanner Pinot noir -- a 30-year history". *Santa Maria Sun*. Retrieved 11 December 2012.

[2] Gaffney, Rusty. "Lane Tanner Winery". *PinotFile*. The Prince of Pinot. Retrieved 11 December 2012.

[3] "Lane Tanner". *People and Institutions*. Calwineries. Retrieved 11 December 2012.

[4] Gilbert, Lucia and John. "Lane Tanner". *Winemakers*. Women Winemakers of California. Retrieved 11 December 2012.

[5] Gaffney, Rusty. "Lane Tanner's Enticing Pinots". *PinotFile*. The Prince of Pinot. Retrieved 11 December 2012.

[6] "Sierra Madre Vineyard Hires Lane Tanner as Winemaker". *People News*. Wine Business. Retrieved 11 December 2012.

[7] Virbila, S. Irene. "2006 Lane Tanner Pinot noir". *Wine of the Week*. Los Angeles Times. Retrieved 11 December 2012.

23.4 External links

- Official website

- Lane Tanner's profile on Sierra Madre Vineyard's website

Chapter 24

Păstorel Teodoreanu

Păstorel Teodoreanu, or just **Păstorel** (born **Alexandru Osvald (Al. O.) Teodoreanu**; July 30, 1894 – March 17, 1964), was a Romanian humorist, poet and gastronome, the brother of novelist Ionel Teodoreanu. He worked in many genres, but is best remembered for his parody texts and his epigrams, and less so for his Symbolist verse. His roots planted in the regional culture of Moldavia, which became his main source of literary inspiration, Păstorel was at once an opinionated columnist, famous wine-drinking bohemian, and decorated war hero. He worked with the influential literary magazines of the 1920s, moving between *Gândirea* and *Viaţa Românească*, and cultivated complex relationships with literary opinion-makers such as George Călinescu.

Teodoreanu's career peaked in 1937, when he received one of Romania's most prestigious awards, the National Prize. He was later employed as a World War II propagandist, which caused him to be shunned by Romanian leftists. From 1947, Păstorel was marginalized and closely supervised by the communist regime, making efforts to adapt his style and politics. Beyond this facade conformity, he contributed to the emergence of an underground, largely oral, anti-communist literature.

In 1959, Teodoreanu was apprehended by the communist authorities, and prosecuted in a larger show trial of Romanian intellectual resistants. He spent some two years in prison, and reemerged as a conventional writer. He died shortly after, without having been fully rehabilitated. His work was largely inaccessible to readers until the 1989 Revolution.

24.1 Biography

24.1.1 Early life and World War I service

The Teodoreanu brothers were born to Sofia Muzicescu, wife of the lawyer Osvald Al. Teodoreanu. Sofia was the daughter of Gavril Muzicescu, a famous composer from Moldavia.[1][2] When Păstorel was born, on July 30,

1894, she and her husband were living at Dorohoi. Ionel (*Ioan-Hipolit Teodoreanu*) and Puiuţu (*Laurenţiu Teodoreanu*) were his younger siblings, born after the family had moved to Iaşi, the Moldavian capital city.[1] Osvald's father, Alexandru T. Teodoreanu, had previously served as City Mayor.[3] The Teodoreanus lived in a townhouse just outside Zlataust Church. They were neighbors of poetess Otilia Cazimir[2] and relatives of novelist Mărgărita Miller Verghy.[4]

From 1906, Alexandru Osvald attended the National High School Iaşi.[5] He had a vivid interest in literary activities and, critics note, acquired a solid classical culture.[6] He was friends with a future literary colleague, Demostene Botez, with whom he shared lodging at the boarding school. Years later, in one of his reviews for Botez's books, Teodoreanu confessed that he once used to steal wine from Botez's own carboy.[7]

In 1914, just as World War I broke out elsewhere in Europe, he was undergoing military training at a Moldavian cadet school.[8] Over the following months, Osvald Teodoreanu became known for his support of prolonged neutrality, which set the stage for a minor political scandal.[9] When, in 1916, Romania joined the Entente Powers, Alexandru was mobilized, a Sub-lieutenant in the 24th artillery regiment, Romanian Land Forces.[10] As he recalled, his emotional father accompanied him as far west as the army would allow.[8]

The future writer saw action in the Battle of Transylvania, then withdrew with the defeated armies into besieged Moldavia. His fighting earned him the Star of Romania and the rank of Captain.[8] Also drafted, Puiuţu Teodoreanu died on the front during the battles of 1918.[11]

24.1.2 Debut and *Gândirea* affiliation

In 1919, upon demobilization, Alexandru returned to Iaşi. Like Ionel, he became a contributor to the magazines *Însemnări Literare* and *Crinul*.[12] He took a law degree from Iaşi University, and, in 1920, moved to the opposite corner of Romania, employed by the Turnu Severin Courthouse.[11] He only spent a few months there. Before the end of the year, he relocated to Cluj, where Cezar Petrescu employed him as a staff writer for his literary magazine, *Gândirea*.[11] The group's activity was centered on Cluj's New York Coffeehouse.[13]

Together with another *Gândirea* author, Adrian Maniu, Teodoreanu wrote the fantasy play *Rodia de aur* ("Golden Pomegranate"). It was published by the Moldavian cultural tribune, *Viața Românească*, and staged by the National Theater Iași in late 1920.[14] Some months later, Teodoreanu was co-opted by theatrologist Ion Marin Sadoveanu into the *Poesis* literary salon, whose members militated for modernism.[15]

In short while, Al. O. Teodoreanu became a popular presence in literary circles, and a famous *bon viveur*. The moniker *Păstorel*, candidly accepted by Teodoreanu, was a reference to these drinking habits: he was said to have "tended" (*păstorit*) the rare wines, bringing them to the attention of other culinary experts.[1] His first contribution to food criticism was published by *Flacăra* on December 31, 1921, with the title *Din carnetul unui gastronom* ("From a Gastronome's Notebook").[16] Teodoreanu integrated with the bohemian society in several cities, leaving written records of his drunken dialogues with linguist Alexandru Al. Philippide.[17] At Iași, the Teodoreanus tightened their links with *Viața Românească*, and with novelist Mihail Sadoveanu. A visitor, poet-critic Felix Aderca, reported seeing Păstorel at *Viața Românească*, "plotting" against the National Theater Bucharest, because, unlike the nationalist theatrical companies of Iași, it only rarely staged Romanian plays. Aderca's antagonistic remarks, published in *Sburătorul*, reflected growing tensions between the modernist circles in Bucharest and the cultural conservatives in Iași.[18]

Teodoreanu's only solo work as a playwright was the one-act comedy *V-a venit numirea* ("Your Appointment Has Been Received"), written in 1922.[19] In 1923, he published his "Inscriptions on a Coffeehouse Table" in the satirical magazine *Hiena*, which was edited by *Gândirea* 's Pamfil Șeicaru.[20] While receiving his first accolades as a writer, Păstorel was becoming a sought-after public speaker. Together with *Gândirea* 's other celebrities, he toured the country and gave public readings from his works (1923).[11] He also made an impact with his welcome speech for Crown Princess Ileana and her "Blue Triangle" Association of Christian Women. The address culminated in a polite pun: "I finally understood that the Blue Triangle is not a circle, but a sum of concentric circles, whose center is Mistress Ileana, and whose radius reaches into our hearts."[21]

24.1.3 *Țara Noastră* and *Rodia de aur*

Teodoreanu was also involved in the cultural and political quarrels of postwar Greater Romania, taking the side of newcomers from Transylvania, who criticized the country's antiquated social system in the name of "integral nationalism".[22] In January 1925, he began writing for the Transylvanian review *Țara Noastră* and became, together with Octavian Goga and Alexandru "Ion Gorun" Hodoș, its staff polemicist.[23] In the mid-1920s, Păstorel's satire had found its main victim: Nicolae Iorga, the influential historian, poet and political agitator. According to Goga and Hodoș, Iorga's older brand of nationalism was unduly self-serving, irresponsible, and confusing.[24]

Teodoreanu followed up with satirical pieces, comparing the omnipresence of Iorga "the demigod" with the universal spread of novelty Pink Pills. He also ridiculed Iorga's ambitions in poetry and literary theory: "Mr. Iorga doesn't get how things work, but he is able to persuade many others: he is dangerous."[25] Teodoreanu was courted by the modernist left-wing circles, which were hostile to Iorga's traditionalism, and was a guest writer for a (formerly radical) art magazine, *Contimporanul*.[26]

Păstorel's editorial debut came only later. In 1928, Cartea Românească publishers issued his parody historical novel, titled *Hronicul Măscăriciului Vălătuc* ("The Chronicle of Jester Harrow").[27] It earned him a literary award sponsored by the Romanian Academy.[28] His *Trei fabule* ("Three Fables") were taken up by *Bilete de Papagal*, an experimental literary newspaper managed by poet Tudor Arghezi.[29]

Teodoreanu made frequent appearances in Bucharest, where in 1929 the National Theater, directed by Liviu Rebreanu, staged a new version of *Rodia de aur*.[30] The event brought Păstorel into collision with the modernists: at *Cuvântul*, theatrical reviewer Ion Călugăru ridiculed *Rodia de aur* as a backward, "childish", play.[31] The verdict infuriated Teodoreanu, who, according to press reports, visited Călugăru at his office, and pummeled him in full view. According to *Curentul* daily, he threatened onlookers not to intervene, brandishing a revolver.[31]

At Casa Capșa, where he was residing ca. 1929,[31] Păstorel was involved in another publicized squabble, throwing cakes at a table where Rebreanu sat together with the modernists Camil Baltazar, Ion Theodorescu-Sion and Ilarie Voronca.[32] At the time, the Ilfov County tribunal received a legal complaint from Călugăru, who accused Teodoreanu of assault and repeated death threats. History does not record whether Teodoreanu was ever brought to court.[31] *Contimporanul* also took its distance from Teodoreanu, who received negative reviews in its pages.[33]

24.1.4   1930s

Păstorel returned to food criticism, with chronicles published in *Lumea*, a magazine directed by literary historian George Călinescu, in *Bilete de Papagal*, and in the left-wing review *Facla*.[16] He was involved in the dispute opposing *Viața Românească* mentor Garabet Ibrăileanu to philologist Giorge Pascu, and, in December 1930, published in *Lumea* two scathing articles against the latter.[34] Pascu sued him for damages.[17]

Also in 1930, he joined the National Theater Iași directorial staff. His colleagues were Moldavian intellectuals from the *Viața Românească* group: Sadoveanu, Demostene Botez, Mihail Codreanu, Iorgu Iordan.[35] Like Sadoveanu and Codreanu, he was inducted into the Romanian Freemasonry's Cantemir Lodge.[36] The formal initiation had an embarrassing twist: Teodoreanu turned up inebriated, and, during the qualifying questionnaire, stated that he "damned well pleased" to become a Mason.[37]

The volume *Strofe cu pelin de mai pentru/contra Iorga Neculai* ("Stanzas in May Wormwood for/against Iorga Neculai") was published in 1931, reportedly at the expense of Păstorel's friends and allies, since it had been refused "by all of the nation's publishing houses".[35] However, bibliographies list it as put out by a *Viața Românească* imprint.[29] The book came out just after Iorga had been appointed Prime Minister. According to one anecdote, the person most embarrassed by the *Strofe* was Osvald Teodoreanu, who had been trying to relaunch his public career. Osvald is said to have toured the Iași bookstores on the day *Strofe* came out, purchasing all copies because they could reach the voters.[38]

More officially, Teodoreanu published two sketch story volumes: in 1931, *Mici satisfacții* ("Small Satisfactions") with Cartea Românească; in 1933, with Editura Națională Ciornei—Rosidor, *Un porc de câine* ("A Swine of a Dog").[39] Eventually, Teodoreanu left Moldavia behind, and moved to Bucharest, where he rented a Grivița house.[17] With help from the cultural policy-maker, General Nicolae M. Condiescu,[40] he was employed as a book reviewer for The Royal Foundations Publishing House, under manager Alexandru Rosetti.[41]

He also became a professional food critic for the literary newspaper *Adevărul Literar și Artistic*, with a column he named *Gastronomice* ("Gastronomics"), mixing real and imaginary recipes.[42] It was in Bucharest that he met and befriended Maria Tănase, Romania's leading female vocalist.[43] Still indulging in his pleasures, Teodoreanu was living beyond his means, pestering Călinescu and Cezar Petrescu with requests for loans, and collecting from all his own debtors.[17] Ibrăileanu, who still enjoyed Teodoreanu's capers and appreciated his talent, sent him for re-

view his novel, *Adela*. Păstorel lost and barely recovered the manuscript, then, in his drunken escapades, forgot to review it, delaying its publication.[17]

A collection of Al. O. Teodoreanu's lampoons and essays, of which some were specifically directed against Iorga, saw print in two volumes (1934 and 1935). Published with Editura Națională Ciornei, it carries the title *Tămâie și otravă* ("Frankincense and Poison"), and notably includes Teodoreanu's thoughts on social and cultural policies.[44] The two books were followed in 1935 by another sketch story volume, eponymously titled *Bercu Leibovici*. In its preface, Teodoreanu announced that he refused to even classify this work, leaving classification to "morons and rubberneckers".[45] The following year, the prose collection *Vin și apă* ("Wine and Water") was issued by Editura Cultura Națională.[46] Also in 1936, Teodoreanu contributed the preface to Romania's standard cookbook, assembled by Sanda Marin.[47]

Osvald Teodoreanu and his two living sons participated in the grand reopening of Hanul Ancuței, a roadside tavern in Tupilați, relocated to Bucharest. The other members and guests were literary, artistic and musical celebrities: Arghezi, D. Botez, Cezar Petrescu, Sadoveanu, Cella Delavrancea, George Enescu, Panait Istrati, Milița Petrașcu, Ion Pillat, Nicolae Tonitza, etc.[13] Păstorel tried to reform the establishment into a distinguished wine cellar, and wrote a code of conduct for the visitors.[48] The pub also tried to engender a literary society, dedicated primarily to the reformation of Romanian literature, and, with its profits, financed young talents.[13]

The Hanul Ancuței episode ended when Teodoreanu was diagnosed with liver failure. Sponsored by the Romanian Writers' Society syndicate, he treated his condition at Karlovy Vary, in Czechoslovakia. The experience, which meant cultural isolation and a teetotal's diet, led Teodoreanu to declare himself an enemy of all things Czechoslovak.[17] During his stays in Karlovy Vary, he corresponded with his employer, Rosetti, keeping with the events in Romania, but wondering if Romanians still remembered him.[17]

Păstorel was a recipient of the 1937 National Prize for Prose. The jury comprised other major writers of the day: Rebreanu, Sadoveanu, Cezar Petrescu, Victor Eftimiu.[45] The author took pride in such recognition. In his definition, the National Prize was an endorsement "worth its weight in gold".[28] He impressed the other literati at the celebratory dinner, where he was "dressed to the nines" and drank with moderation.[38] After the event, Teodoreanu turned his attention to his poetry writing: in 1938, he published the booklet *Caiet* ("Notebook").[49] The same year, Ionel joined his older brother in Bucharest.[50]

24.1.5 World War II propagandist

The Teodoreanu brothers were public supporters of the authoritarian regime instituted, in 1938, by King Carol II, contributing to the government propaganda.[51] The king returned the favor and, also in 1938, Păstorel was made a Knight of *Meritul Cultural* Order, 2nd Class.[52] From autumn 1939, when the start of World War II left Romania exposed to a foreign invasion, Teodoreanu was again called under arms. Stationed with his 24th artillery regiment in the garrison of Roman,[53] he put on hold his regular food chronicles.[42] However, his military duties quickly dissolved into wine-drinking meals. This was attested by Corporal Gheorghe Jurgea-Negrilești, an aristocrat and memoirist, who served under Teodoreanu and remained his friend in civilian life.[37]

In 1940, Teodoreanu worked with Ion Valentin Anestin, writing the editorial "Foreword" to Anestin's satirical review, *Gluma*, and published a series of aphorisms in *Revista Fundațiilor Regale*.[54] The writer was living in Bucharest, at the Carlton Hotel. The building was destroyed in the November 10 earthquake, and, for a while, Teodoreanu himself was presumed dead.[55]

By then, Romania, under *Conducător* Ion Antonescu, became an ally of Nazi Germany. In summer 1941, the country joined in the German attack on the Soviet Union (Operation Barbarossa). Teodoreanu took employment as an Antonescu regime propagandist, publishing, in the newspaper *Universul*, a panegyric dedicated to pilot Horia Agarici.[56] *Țara* newspaper of Sibiu hosted his scathing anti-communist poem, *Scrisoare lui Stalin* ("A Letter to Stalin").[57] A second edition of *Bercu Leibovici* came out in 1942,[53] followed in 1934 by a reprint of *Caiet*.[58]

Still living in Bucharest, Teodoreanu kept company with Jurgea-Negrilești. According to the latter, Păstorel had friendly contacts with novelist Paul Morand, who was the diplomatic representative of Vichy France in Bucharest. The story shows a high-strung Teodoreanu, who defied wartime restriction to obtain a bowler hat and gloves, and dressed up for one of Morand's house-parties.[59] In mid-1944, at the peak of Allied bombing raids, Teodoreanu had taken refuge in Budești, a rural commune south of the capital. He was joined there by Maria Tănase and her husband of the time.[43]

After the King Michael's Coup broke apart Romania's alliance with the Axis Powers, Teodoreanu returned to regular journalism. His food criticism was again taken up by *Lumea*, and then by the general-interest *Magazin*.[60] Lacking a stable home, he was hosted at The Royal Foundations Publishing House, and could be seen walking about its library in a red housecoat.[61] Teodoreanu's contribution to wartime propaganda made him a target for retribution in the Romanian Communist Party press. Already in October 1944, *România Liberă* and *Scînteia* demanded for him to be excluded from the Writers' Society, noting that he had "written in support of the anti-Soviet war".[62]

24.1.6 Communist takeover

The restaurant section of a Romanian consumer cooperative, 1950

Păstorel's career was damaged by the full imposition, in 1947, of a Romanian communist regime. In May 1940, Teodoreanu had defined humor as "the coded language that smart people use to understand each other under the fools' noses".[63] Resuming his food writing after 1944, he began inserting subtle jokes about the new living conditions, even noting that the widespread practice of rationing made his texts seem "absurd".[64] Traditionally, his cooking recommendations had been excessive, and recognized as such by his peers. He firmly believed that *cozonac* cake required 50 eggs for each kilogram of flour (that is, some 21 per pound).[48]

The communists were perplexed by the Teodoreanu case, undecided about whether to punish him as a dissident or enlist him as a fellow traveler.[65] Păstorel was experiencing financial ruin, living on commissions, handouts and borrowings. He tried to talk Maria Tănase into using his poems as song lyrics, and stopped seeing her altogether when her husband refused to lend him money.[43] His brother Ionel died suddenly in February 1954, leaving Păstorel devastated.[53] He compensated for the loss by keeping company with other intellectuals of the anti-communist persuasion. His literary circle, hosted by the surviving Bucharest locales, included, among others, Jurgea-Negrilești, Şerban Cioculescu, Vladimir Streinu, Aurelian Bentoiu, and Alexandru Paleologu.[66]

By 1954, Teodoreanu was being called in for questioning by agents of the Securitate, the communist secret police. Pressure was put on him to divulge his friends' true feelings about the political regime. He avoided a direct answer, but eventually informed Securitate about Maria Tănase's apparent disloyalty.[43] While harassed in this manner, Teodoreanu was already earning a leading place in underground counterculture, where he began circulating his new anti-communist compositions. According to literary critic Ion Simuț, the clandestine poetry of Păstorel, Vasile Voiculescu and Radu Gyr is the only explicit negation of communism to have emerged from 1950s Romania.[67] As other Securitate records show, the public was aware of Teodoreanu's visits to the Securitate, but distinguished between him, who was "called over" to confess, and those who made voluntary denunciations.[68]

In trying to salvage his career, Teodoreanu was forced to diversify his literary work. In 1956, his literary advice for debuting authors was hosted by the gazette *Tînărul Scriitor*, an imprint of the Communist Party School of Literature.[69] He also completed and published translations from Jaroslav Hašek (*Soldier Švejk*) and Nikolai Gogol (*Taras Bulba*).[53] In 1957, he prefaced the collected sonnets of Mihail Codreanu,[53] and issued, with Editura Tineretului, a selection of his own prose, *Berzele din Boureni* ("The Storks of Boureni").[70] With Călinescu, Teodoreanu worked on *La Roumanie Nouvelle*, the French-language communist paper, where he had the column *Goutons voir si le vin est bon* ("Let's Taste the Wine and See if It's Good").[71]

From 1957 to 1959, Teodoreanu resumed his food chronicles in *Magazin*, while also contributing culinary reviews in *Glasul Patriei* and other such communist propaganda newspapers.[72] According to researcher Florina Pîrjol: "the scion of bourgeois intellectuals, with his liberal values and his aristocratic spirit, unsuitable for political 'taming', Al. O. Teodoreanu had a rude awakening into a world where, perceived as a hostile element, he was unable to exercise his profession".[73] According to literary reviewer G. Pienescu, who worked with Teodoreanu in the 1960s, the *Glasul Patriei* collaboration was supposed to grant Păstorel a "certificate of good citizenship".[61]

24.1.7   Censorship and show trial

Under pressure from communist censorship, Teodoreanu was reconfiguring his literary profile. Dropping all references to Western cuisine, his food criticism became vague, reusing agitprop slogans about "goodwill among men", before adopting in full the communists' wooden tongue.[74] Although the country was still undernourished, Păstorel celebrated the public self-service chain, *Alimentara*, as a "structural transformation" of the Romanian psyche.[75] Meanwhile, some anti-communist texts, circulated by Teodoreanu among the underground dissidents, were intercepted by the authorities. Those who have documented Teodoreanu's role in the development of un-

derground humor note that he paid a dear price for his contributions.[67][76] On October 30, 1959, he was arrested,[73] amidst a search for incriminating evidence. The Securitate relied on reports from its other informers, one of whom was Constantin I. Botez, the psychologist and academic.[68]

The writer became one of 23 intellectuals implicated in a show trial, whose main victims were Constantin Noica and Dinu Pillat. Although grouped together, these men and women were accused of a variety of seditious deeds, from engaging in "hostile conversations" to keeping company with Western visitors.[77] One thing they had in common was their relationship with Noica: they had all attended meetings in Noica's home, listening to his readings from the letters of a banished philosopher, Emil Cioran.[78]

Teodoreanu received a sentence of six years in "correctional prison", with three years of loss of rights, and permanent confiscation of his assets.[73] Communist censors took over his manuscripts, some of which were unceremoniously burned.[73] These circumstances forced Teodoreanu's wife, Marta, to work nights as a street sweeper.[79]

Held in confinement at Gherla prison, Teodoreanu filed an appeal. He admitted to having ridiculed communism, and to having distanced himself from Socialist Realism, but asked to be allowed a second chance, stating his usefulness in writing "propaganda".[73] Reportedly, the Writers' Union, whose President was Demostene Botez, made repeated efforts to obtain his liberation. Teodoreanu was not informed of this, and was shocked to encounter Botez, come to plead in his favor, in the prison warden's office.[37] He was ultimately granted a reprieve in April 1962, together with many other political prisoners, and allowed to return to Bucharest.[66]

24.1.8   Illness and death

Teodoreanu returned to public life, but was left without the right of signature, and was unable to support himself and Marta. In this context, he sent a letter to the communist propaganda chief, Leonte Răutu, indicating that he had "redeemed his past", and asking to be allowed back into the literary business.[80] Păstorel made his comeback with the occasional column, in which he continued to depict Romania as a land of plenty. Written for Romanian diaspora readers, just shortly after the peak of food restrictions, these claimed that luxury items (Emmental, liverwurst, Nescafé, Sibiu sausages) had been made available in every neighborhood shop.[81] His hangout was the Ambasador Hotel, where he befriended an eccentric communist, poet Nicolae Labiș.[82] Helped by Pienescu, he was preparing a collected works edition, *Scrieri* ("Writings"). The communist censors were adverse to its publishing, but, after Tudor Arghezi spoke in Teodoreanu's favor, the book was included in the "fit for publishing" list of 1964.[61]

Păstorel was entering the terminal stages of lung cancer, receiving palliative care at his house on Vasile Lascăr Street, Bucharest.[61] Teodoreanu's friend and biographer, Alexandru Paleologu, calls his "an exemplary death". According to Paleologu, Teodoreanu had taken special care to render his suffering bearable for those around him, being "lucid and courteous".[53] Jurgea-Negrilești was present at one of the group's last meetings, recalling: "At the very last drop [of wine], he got up on his feet... there was a gravity about him, a greatness that I find hard to explain. In a voice that his pain had made hoarse, he asked that we leave him alone".[83]

Teodoreanu died at home, on March 17, 1964, just a day after Pienescu brought him news that censorship had been bypassed;[61] in some sources, the date of death is given as March 15.[53] He was buried, alongside Ionel Teodoreanu, at Bellu cemetery.[84] Owing to Securitate surveillance, the funeral was a quiet affair. The Writers' Union was only represented by two former *Gândirea* contributors, Maniu and Nichifor Crainic. They were not mandated to speak about the deceased, and kept silent, as did the Orthodox Church priest who was supposed to deliver the service.[85]

The writer had left two translations (Anatole France's *Chronicle of Our Own Times*; Prosper Mérimée's *Nouvelles*), first published in 1957.[53] As Pienescu notes, he had never managed to sign the contract for *Scrieri*.[61]

24.2 Work

24.2.1 Jester Harrow

Culturally, Teodoreanu belonged to the schools of interwar nationalism, be they conservative (*Gândirea*, *Țara Noastră*) or progressive (*Viața Românească*). Some exegetes have decoded proof of patriotic attachment in the writer's defense of Romanian cuisine, and especially his ideas about Romanian wine. Șerban Cioculescu once described his friend as a "wine nationalist"[86] and George Călinescu suggested that Păstorel was entirely out of his element when discussing French wine.[87] On one hand, Păstorel supported illusory claims of Romanian precedence (including a story that caviar was discovered in Romania); on the other, he issued loving, if condescending, remarks about Romanians being a people of "grill cooks and *mămăligă* eaters".[71] However, Teodoreanu was irritated by the contemplative traditionalism of Moldavian writers, and, as Cioculescu writes, his vitality clashed with the older schools of nationalism: Nicolae Iorga's *Sămănătorul* circle and "its Moldavian pair", Poporanism.[88] Philosophically, he remained indebted to Oscar Wilde and the aestheticists.[89]

The frame story *Hronicul Măscăriciului Vălătuc* is remembered as a most atypical contribution to Romanian literature, and, critics argue, "one of his most valuable books",[90] a "masterpiece".[91] Nevertheless, the only commentator to have been impressed by the totality of *Hronicul*, and to have rated Păstorel as one of Romania's greatest humorists, is essayist Paul Zarifopol. His assessment was challenged, even ridiculed, by the academic community.[92] The consensus is nuanced by critic Bogdan Crețu, who writes: "Păstorel may well be, as far as some care to imagine, peripheral in literature, but [...] he is not at all a minor writer."[17]

According to Călinescu, *Hronicul Măscăriciului Vălătuc* parallels Balzac's *Contes drôlatiques*. Like the *Contes*, Jester Harrow's tale reuses, and downgrades, the conventions of medieval historiography—in Păstorel's text, the material for parody is Ion Neculce's *Letopisețul țărâi Moldovei*.[93] As Călinescu notes, Teodoreanu mixed the subversive "counterfeiting" of Neculce's history into his own homage to the verbal clichés of Moldavian dialects.[94] In a 1929 interview, Păstorel specified his models: the Moldavian chroniclers, Neculce and Miron Costin; the modern pastiches, Balzac's *Contes* and Anatole France's *Merrie Tales of Jaques Tournebroche*.[95] Literary historian Eugen Lovinescu believed that Teodoreanu was naturally closer to the common source of these works, namely the fantasy stories of François Rabelais. Păstorel's "so very Rabelaisian" writing has a "thick, big, succulent note, that will saturate and overfill the reader."[96]

A narrative experiment, *Hronicul* comprises at least five parody "historical novels", independent of each other: *Spovedania Iancului* ("Iancu's Confession"), *Inelul Marghioliței* ("Marghiolița's Ring"), *Pursângele căpitanului* ("The Captain's Purebred"), *Cumplitul Trașcă Drăculescul* ("Trașcă the Terrible, of the Dracula Clan"), and *Neobositulŭ Kostakelŭ* ("Koštakel ye Tirelefs"). In several editions, they are bound together with various other works, covering several literary genres. According to biographer Gheorghe Hrimiuc, the latter category is less accomplished than the "chronicle".[97] It notably includes various of Teodoreanu's attacks on Iorga.[98]

Although the presence of anachronisms makes it hard to even locate the stories, they seem to be generally referencing the 18th- and 19th-century Phanariote era, during which Romanians adopted a decadent, essentially anti-heroic, lifestyle.[99] A recurrent theme is that of the colossal banquet, in most cases prompted by nothing other than the joy of company or a *carpe diem* mentality, but so excessive that they drive the organizers into moral and material bankruptcy.[100] In all five episodes, Păstorel disguises himself as various unreliable narrators. He is, for instance, a decrepit General Coban (*Pursângele căpitanului*) and a retired courtesan (*Inelul Marghioliței*). In *Neobositulŭ Kostakelŭ*, a "found manuscript", he has three narrative voices: that of the writer, Pantele; that of the skeptic reviewer, Balaban; and that of the concerned "philologist", with his absurd critical apparatus (a parody of scientific conventions).[101] The alter ego, "Harrow", is only present (and mentioned by name) in the rhyming *Predoslovie* ("Foreword"), but is implicit in all the stories.[102]

Also in *Neobositulŭ Kostakelŭ*, Teodoreanu's love for role-playing becomes a study in intertextuality and candid stupidity. Pantele is a reader of Miron Costin, but seemingly incapable of understanding his literary devices. He reifies metaphoric accounts about Moldavia "flowing with milk and honey": "Had this been in any way true true, people would be glued to fences, like flies".[103] Even the protagonist, Kostakel, is a writer, humorist and parodist, who has produced his own chronicle of "obscenities" with the stated purpose of irritating Ion Neculce (who thus makes a brief appearance *within* Harrow's "chronicle").[104] The deadpan critical apparatus accompanying such intertextual dialogues is there to divert attention from Teodoreanu's narrative tricks and anachronisms. Hrimiuc suggests that, by pretending to read his own "chronicle" as a valid historical record, Păstorel was sending in "negative messages about how not to decode the work."[105]

Neobositulŭ Kostakelŭ and *Pursângele căpitanului* comprise some of Păstorel's ideas about the Moldavian ethos. The locals have developed a strange mystical tradition, worshiping Cotnari wine, and regarding those who abstain from it as "enemies of the church".[106] In *Neobositulŭ Kostakelŭ*, the antagonist is Panagake, an outsider (a Graeco-Romanian) and usurper of tradition. Although he suffers defeat and ridicule, he is there to announce that the era of sheer merrymaking is about to end.[107] As critic Doris Mironescu notes, the characters experience a "entry into time", except "theirs is not Great history, but a minor one, that of intimate disasters, of homemaking tragedies and the domestic hell."[108]

Hronicul satirizes the conventions of Romanian neoromanticism and of the commercial adventure novel, particularly so in *Cumplitul Trașcă Drăculescul*.[109] The eponymous hero is a colossal and unpredictable hajduk, born with the necessary tragic flaw. He lives in continuous erotic frenzy, pushing himself on all available women, "without regard as to whether they were virgins or ripe women, not even if they had happened to be his cousins or his aunts".[110] Still, he is consumed by his passion for the nubile Sanda, but she dies, of "chest trouble", on the very night of their wedding. The broken Trașcă commits suicide on the spot. These events are narrated with the crescendo of romantic novels, leading to the unceremonious punch line: "And it so happened that this Trașcă of the Draculas was ninety years of age."[111]

24.2.2　Caragialesque prose

Teodoreanu's *Mici satisfacții* and *Un porc de câine* echo the classical sketch stories of Ion Luca Caragiale, a standard in Romanian humor. Like him, Păstorel looks into the puny lives and "small satisfactions" of Romania's *petite bourgeoisie*, but does not display either Caragiale's malice or his political agenda.[17][112] His own specialty is the open-ended, unreliably-narrated, depiction of mundane events: the apparent suicide of a lapdog, or (in *Berzele din Boureni*) an "abstruse" dispute about the flight patterns of storks.[113]

Un porc de câine pushed the jokes a little further, risking to be branded an obscene work. According to critic Perpessicius, "a witty writer can never be an obscene writer", and Păstorel had enough talent to stay out of the pornographic range.[114] Similarly, Cioculescu describes his friend as an artisan of "libertine humor", adverse to didactic art, and interested only in "pure comedy".[115] In his narrator's voice, Păstorel mockingly complains that the banal was being replaced by the outstanding, making it hard for humorists to find subject matters. Such doubts are dispelled by the intrusion of a blunt, but inspirational, topic: "Can it be true that mayweed is an aphrodisiac?"[116] In fact, *Un porc de câine* expands Teodoreanu's range beyond the everyday, namely by showing the calamitous, entirely unforeseeable, effects of an erotic farce.[117] The volume also includes a faux obituary, honoring the memory of one Nae Vasilescu. This stuttering tragedian, whose unredeemed ambition was to play Shylock, took his revenge on the acting profession by becoming a real-life usurer—an efficient if dishonorable way to earning the actors' fear and respect.[118]

Critics have rated Teodoreanu as a Caragialesque writer, or a "Moldavian", "thicker", more archaic Caragiale.[17][119] Hrimiuc suggests that Caragiale has become an "obligatory" benchmark for Teodoreanu's prose, with enough differences to prevent Păstorel from seeming an "epigone".[120] Hrimiuc then notes that Teodoreanu is entirely himself in the sketch *S-au supărat profesorii* ("The Professors Are Upset"), fictionalizing the birth of the National Liberal Party-Brătianu with "mock dramaticism", and in fact poking fun at the vague political ambitions of Moldavian academics.[121]

As a Caragiale follower, Teodoreanu remained firmly within established genre. Doris Mironescu describes his enrollment as a flaw, placing him in the vicinity of "minor" Moldavian writers (I. I. Mironescu, Dimitrie D. Pătrășcanu).[122] The other main influence, as pinpointed by the literary critics, remains Teodoreanu's personal hero, Anatole France.[123] In *Tămâie și otravă*, Teodoreanu is, like France, a moralist. However, Călinescu notes, he remains a "jovial" and "tolerable" one.[124]

24.2.3 Symbolist poetry

Păstorel had very specific tastes in poetry, and was an avid reader of the first-generation Symbolists. Of all Symbolist poets, his favorite was Paul Verlaine,[125] whose poems he had memorized to perfection,[38][126] but he also imitated Henri de Régnier, Albert Samain and Jean Richepin.[127] Like Verlaine, Teodoreanu had mastered classical prosody, so much so that he believed it was easier, and more vulgar, for one to write in verse—overall, he preferred prose.[128] He was entirely adverse to Romania's modernist literature, most notably so when he ridiculed the poetry of Camil Baltazar.[7]

In *Caiet*, Teodoreanu is a poet of the macabre, honoring the ghoulish genre invented by his Romanian Symbolist predecessors. According to critics such as Călinescu[129] and Alexandru Paleologu, his main reference is Alexandru Macedonski, the Romanian Symbolist master. Paleologu notes that Păstorel is the more "lucid" answer to Macedonski's unlimited "Quixotism".[130] Together with the *carpe diem* invitation in *Hronicul*, *Caiet* is an implicit celebration of life:

Teodoreanu's contribution to Romanian poetry centers on an original series, *Cântecèle de ospiciu* ("Tiny Songs from a Hospice"), written from the perspective of the dangerously insane. As Călinescu notes, they require "subtle humor" from the reader.[129] They include delirious monologues:

24.2.4 Scattered texts and apocrypha

As a poet of the mundane, Teodoreanu shared glory with the other *Viața Românească* humorist, George Topîrceanu. If their jokes had the same brevity,[122] their humor was essentially different, in that Topîrceanu preserved an innocent worldview.[131] In this class of poetry, Teodoreanu had a noted preference for orality, and, according to interwar essayist Petru Comarnescu, was one of Romania's "semifailed intellectuals", loquacious and improvident.[132] As an impish journalist, he always favored the ephemeral.[133] Păstorel's work therefore includes many scattered texts, some of which were never collected for print. Gheorghe Hrimiuc assessed that his aphorisms, "inscriptions" and self-titled "banal paradoxes" must number in the dozens, while his epigram production was "enormous".[134]

In his attacks on Nicolae Iorga, the epigrammatist Păstorel took the voice of Dante Aligheri, about whom Iorga had written a play. Teodoreanu's Dante addressed his Romanian reviver, and kindly asked to be left alone.[135] Anti-Iorga epigrams abound in *Țara Noastră* pages. Attributable to Teodoreanu, they are signed with various irreverent pen names, all of them referencing Iorga's various activities and opinions: Iorgu Arghiropol-Buzatu, Hidalgo Bărbulescu,

Mița Cursista, Nicu Modestie, Mic dela Pirandola.[136] On the friendly side, the fashion of exchanging epigrams was also employed by Teodoreanu and his acquaintances. In one such jousting, with philosopher Constantin Noica, Teodoreanu was ridiculed for overusing the apostrophe (and abbreviation) to regulate his prosody; Teodoreanu conceded that he could learn "writing from Noica".[137]

Other short poems merely address the facts of life in Iași or Bucharest. His first ever quatrain, published in *Crinul*, poked fun at the Imperial Russian Army, whose soldiers were still stationed in Moldavia.[12] A later epigram locates the hotspot of prostitution in Bucharest: the "maidens" of Popa Nan Street, he writes, "are beautiful, but they're no maidens".[138] In 1926, *Contimporanul* published his French-language calligram and "sonnet", which recorded in writing a couple's disjointed replies during the sexual act.[33]

In Călinescu's opinion, these works should be dismissed. They are, he notes, "without spirit", "written in a state of excessive joy, that confuses the writer about the actual suggestive power of his words."[48] Teodoreanu's artistic flair was poured into his regular letters, which fictionalize, rhyme and dramatize everyday occurrences. These texts "push into the borders of literature" (Hrimiuc),[125] and are worthy of a "list of great epistolaries" (Crețu).[17]

Urban folklore and communist prosecutors recorded a wide array of anti-communist epigrams, attributed (in some cases, dubiously)[66][76] to Al. O. Teodoreanu. He is the purported author of licentious comments about communist writer Veronica Porumbacu and her vagina,[139] and about the "arselicking" communist associate, Petru Groza.[47][76] Tradition also credits him with the corrosive joke about the Statue of the Soviet Liberator, a monument which towered over Bucharest from 1946:

Elsewhere, Teodoreanu derided the communists' practice of enrolling former members of the fascist Iron Guard, nominal enemies, into their own Workers' Party. His unflattering verdict on this unexpected fusion of the political extremes was mirrored by co-defendant Dinu Pillat, in the novel *Waiting for the Last Hour*.[140] Teodoreanu's famous stanza is implicitly addressed to "Captain" Corneliu Zelea Codreanu, the Guard's founder and patron saint:

24.3 In cultural memory

With his constant networking, Păstorel Teodoreanu made a notable impact in the careers of other writers, and, indirectly, on visual arts. Some of his works came with original drawings: illustrations by Ion Sava (for *Strofe cu pelin de mai*);[38] a portrait of the writer, by Ștefan Dimitrescu (*Mici satisfacții*); and graphics by Ion Valentin Anestin (*Vin și apă*.[141] One of the first to borrow from *Hronicul* was George Lesnea, the author of humorous poems about Moldavia's distant past,[142] and a recipient of the Hanul Ancuței literary prize.[13] A young author of the 1940s, Ştefan Baciu, drew inspiration from both *Gastronomice* and *Caiet* in his own humorous verse.[143]

In the 1970s, when liberalization touched Romanian communism, most restrictions on Teodoreanu's work were lifted. Already in 1969, a retrial had cleared the path for his rehabilitation.[47] 1972 was a breakthrough year in his recovery, with a selection of his poems and a new edition of *Hronicul*; the latter was to become "the most readily reedited" Teodoreanu work, down to 1989.[144] Later years brought a bibliophile edition of his *Gastronomice*, with drawings by Done Stan, and a selection of food criticism, *De re culinaria* ("On Food").[145]

Since 1975, Iași has hosted an epigrammatists' circle honoring Teodoreanu's memory. Known as "Păstorel's Free Academy", it originally functioned in connection with *Flacăra Iașului* newspaper, and was therefore kept in check by the communist authorities.[146] In 1988, at Editura Sport-Turism, critic Mircea Handoca published a travel account and literary monograph: *Pe urmele lui Al. O. Teodoreanu-Păstorel* ("On the Trail of Al. O. Teodoreanu-Păstorel").[17][147]

After the Romanian Revolution of 1989 lifted communist restrictions, it became possible for exegetes to investigate the totality of Teodoreanu's contributions. The anti-communist apocrypha have been featured in a topical volume, edited by Gheorghe Zarafu and Victor Frunză in 1996, but remain excluded from the standard Teodoreanu collections (including one published by Rodica Pandele at Humanitas).[67] Under the new regime, food writing was again a profession, and Păstorel became a direct inspiration for gastronomes such as Radu Anton Roman or Bogdan Ulmu, who wrote "*à la Păstorel*".[16]

As such, Doris Mironescu suggests, Teodoreanu made it into "a *sui-generis* national pantheon" of epigrammatists, with Lesnea, Cincinat Pavelescu, and Mircea Ionescu-Quintus.[122] Formal public recognition came in 1997, when the Museum of Romanian Literature honored the Teodoreanu brothers' memory with a plaque, unveiled at their childhood home in Iași.[50] However, the building was partly demolished by its new owners in 2010, a matter which fueled political controversies.[2][3]

24.4 Notes

[1] Teodoreanu & Ruja, p.7

[2] (Romanian) Vasile Iancu, "Memoria culturală, prin grele pătimiri", in *Convorbiri Literare*, May 2011

[3] (Romanian) Gina Popa, "Se stinge 'ulița copilăriei' ", in *Evenimentul*, March 31, 2010

[4] (Romanian) Elena Cojuhari, "Viața și activitatea Margaretei Miller-Verghy în documentele Arhivei Istorice a Bibliotecii Naționale a României", in *Revista BNR*, Nr. 1-2/2009, p.46, 62

[5] Teodoreanu & Ruja, p.7-8

[6] Călinescu, p.777; Hrimiuc, p.293, 295-296; Teodoreanu & Ruja, p.13

[7] Călinescu, p.777

[8] Teodoreanu & Ruja, p.8

[9] Lucian Boia, *"Germanofilii". Elita intelectuală românească în anii Primului Război Mondial*, Humanitas, Bucharest, 2010, p.95. ISBN 978-973-50-2635-6

[10] Teodoreanu & Ruja, p.8-9

[11] Teodoreanu & Ruja, p.9

[12] Tudor Opriș, *Istoria debutului literar al scriitorilor români în timpul școlii (1820-2000)*, Aramis Print, Bucharest, 2002, p.135. ISBN 973-8294-72-X

[13] (Romanian) Constantin Coroiu, "Mitul cafenelei literare", in *Cultura*, Nr. 302, December 2010

[14] Teodoreanu & Ruja, p.9. See also Călinescu, p.1020, 1022; Lovinescu, p.304

[15] Cernat (2007), p.270-271

[16] Pîrjol, p.19, 25

[17] (Romanian) Bogdan Crețu, "Corespondența lui Păstorel", in *Ziarul Financiar*, October 22, 2009

[18] Piru, p.128

[19] Hrimiuc, p.292

[20] Hrimiuc, p.333

[21] Călinescu, p.777-778

[22] Ghemeș, p.68

[23] Ghemeș, p.67, 69

[24] Ghemeș, p.69, 70-72

[25] Ghemeș, p.69-70

[26] Cernat (2007), p.151-152

[27] Călinescu, p.1020; Hrimiuc, p.292, 298; Teodoreanu & Ruja, p.9

[28] Hrimiuc, p.295

[29] Călinescu, p.1020

[30] Călinescu, p.1022; Teodoreanu & Ruja, p.10

[31] (Romanian) Dumitru Hîncu, "Acum optzeci de ani - Bătaie la *Cuvântul*", in *România Literară*, Nr. 44/2009

[32] (Romanian) Daniela Cârlea şontică, "La un şvarţ cu capşiştii", in *Jurnalul Naţional*, August 28, 2006

[33] Cernat (2007), p.152

[34] Piru, p.160, 189

[35] Teodoreanu & Ruja, p.10

[36] (Romanian) Ion Simuţ, "Sadoveanu francmason", in *România Literară*, Nr. 10/2008

[37] (Romanian) Constantin Ţoiu, "Întâmplări cu Păstorel", in *România Literară*, Nr. 51-52/2008

[38] (Romanian) Rodica Mandache, "Boema. La Capşa cu Ion Barbu, Păstorel, Şerban Cioculescu", in *Jurnalul Naţional*, May 12, 2012

[39] Călinescu, p.1020; Hrimiuc, p.292; Teodoreanu & Ruja, p.10-11

[40] Boia (2012), p.114

[41] Teodoreanu & Ruja, p.13

[42] Pîrjol, p.19-20

[43] (Romanian) "Păstorel toarnă la Securitate", in *Jurnalul Naţional*, June 25, 2007

[44] Teodoreanu & Ruja, p.5-6, 11-13

[45] Teodoreanu & Ruja, p.14

[46] Călinescu, p.1020; Hrimiuc, p.292; Teodoreanu & Ruja, p.14

[47] Pîrjol, p.25

[48] Călinescu, p.778

[49] Hrimiuc, p.292; Teodoreanu & Ruja, p.14

[50] (Romanian) Constantin Ostap, "Ionel Teodoreanu, 50 de ani de la moarte", in *Convorbiri Literare*, December 2004

[51] Boia (2012), p.126, 142, 148-149, 167

[52] Boia (2012), p.127

[53] Teodoreanu & Ruja, p.15

[54] Hrimiuc, p.333, 334. See also Popa, p.91

[55] (Romanian) Simona Vasilache, "Dovezi de admiraţie", in *România Literară*, Nr. 28/2009

[56] (Romanian) Lucian Vasile, "Manipularea din presă în prima lună din al doilea război mondial", in *Historia*, April 2011

[57] (Romanian) Monica Grosu, "Din tainele arhivelor", in *Luceafărul*, Nr. 15/2011

[58] Hrimiuc, p.292, 334

[59] (Romanian) Cosmin Ciotloş, "Memorie versus memorialistică", in *România Literară*, Nr. 6/2008

[60] Pîrjol, p.20

[61] (Romanian) G. Pienescu, "Al. O. Teodoreanu", in *România Literară*, Nr. 27/2007

[62] Victor Frunză, *Istoria stalinismului în România*, Humanitas, Bucharest, 1990, p.251, 565. ISBN 973-28-0177-8

[63] Hrimiuc, p.302

[64] Pîrjol, p.20, 25

[65] Pîrjol, p.18-19

[66] Pîrjol, p.21, 25

[67] (Romanian) Ion Simuţ, "A existat disidenţă înainte de Paul Goma?", in *România Literară*, Nr. 22/2008

[68] (Romanian) Adrian Neculau, "O zi din viaţa lui Conu Sache", in *Ziarul de Iaşi*, November 6, 2010

[69] (Romanian) Paul Cernat, "Anii '50 şi *Tînărul Scriitor*", in *Observator Cultural*, Nr. 285, August 2005

[70] Hrimiuc, p.333; Pîrjol, p.22

[71] Pîrjol, p.22

[72] Pîrjol, p.20-21, 22, 24, 26

[73] Pîrjol, p.21

[74] Pîrjol, p.22-24

[75] Pîrjol, p.23

[76] (Romanian) "Gheorghe Grigurcu în dialog cu Şerban Foarţă", in *România Literară*, Nr. 51-52/2007

[77] (Romanian) Alex. Ştefănescu, "Scriitori arestaţi (1944-1964)", in *România Literară*, Nr. 23/2005

[78] (Romanian) Gabriel Liiceanu, "Spovedania lui Steinhardt", in *Dilemateca*, Nr. 1, May 2006 (republished by *România Culturală*). See also Boia (2012), p.280

[79] (Romanian) Al. Săndulescu, "Al doilea cerc", in *România Literară*, Nr. 37/2006

[80] Pîrjol, p.21-22

[81] Pîrjol, p.24

[82] (Romanian) Constantin Ţoiu, "Păstorel recomandă: piftie de cocoş bătrân", in *România Literară*, Nr. 51-52/2006

[83] (Romanian) Paul Cernat, "Senzaţionalul unor amintiri de mare clasă", in *Observator Cultural*, Nr. 130, August 2002

[84] Gheorghe G. Bezviconi, *Necropola Capitalei*, Nicolae Iorga Institute of History, Bucharest, 1972, p.269; (Romanian) Constantin Ostap, "Păstorel Teodoreanu, reeditat în 2007", in *Ziarul de Iași*, February 7, 2007

[85] (Romanian) Ion Constantin, *Pantelimon Halippa neînfricat pentru Basarabia*, Editura Biblioteca Bucureștilor, Bucharest, 2009, p.181. ISBN 978-973-8369-64-1

[86] Hrimiuc, p.327; Mironescu, p.16

[87] Călinescu, p.776

[88] Hrimiuc, p.320-321

[89] Hrimiuc, p.297-298; Mironescu, p.16

[90] Pîrjol, p.19

[91] Hrimiuc, p.295, 311

[92] (Romanian) Alex. Cistelecan, "Paul Zarifopol, partizanul 'adevărului critic integral' ", in *Cultura*, Nr. 388, February 2011; Andreea Grinea Mironescu, "Locul lui Paul Zarifopol. Note din dosarul receptării critice", in *Timpul*, Nr. 10/2011, p.8, 9

[93] Călinescu, p.776; Hrimiuc, p.317; Teodoreanu & Ruja, p.9-10

[94] Călinescu, p.776; Teodoreanu & Ruja, p.10

[95] Hrimiuc, p.317; Mironescu, p.16

[96] Lovinescu, p.208

[97] Hrimiuc, p.311-312

[98] Ghemeș, p.75

[99] Hrimiuc, p.316-317, 325-326; Mironescu, p.16, 17

[100] Hrimiuc, p.321-326, 330-332; Mironescu, p.17

[101] Hrimiuc, p.312-316, 321-322, 329-331; Mironescu, *passim*

[102] Hrimiuc, p.321-322; Mironescu, p.16

[103] Hrimiuc, p.313-315

[104] Hrimiuc, p.322

[105] Hrimiuc, p.315-316

[106] Hrimiuc, p.326-328

[107] Hrimiuc, p.325-326

[108] Mironescu, p.17

[109] Hrimiuc, p.316, 317-321, 330

[110] Hrimiuc, p.325

[111] Hrimiuc, p.318. See also Lovinescu, p.208; Mironescu, p.17

[112] Hrimiuc, p.296-301; Teodoreanu & Ruja, p.10-11

[113] Hrimiuc, p.302-304

[114] Teodoreanu & Ruja, p.11

[115] Hrimiuc, p.308

[116] Hrimiuc, p.306-307

[117] Hrimiuc, p.305-306

[118] Hrimiuc, p.308-310

[119] Călinescu, p.776-777; Teodoreanu & Ruja, p.13

[120] Hrimiuc, p.296-297, 300-301

[121] Hrimiuc, p.297, 299-300

[122] Mironescu, p.16

[123] Hrimiuc, p.295-296; Pîrjol, p.20

[124] Teodoreanu & Ruja, p.11-12

[125] Hrimiuc, p.293

[126] (Romanian) Al. Săndulescu, "Mâncătorul de cărți", in *România Literară*, Nr. 11/2008

[127] Călinescu, p.778, 779

[128] Hrimiuc, p.293-295

[129] Călinescu, p.779

[130] Teodoreanu & Ruja, p.13-14

[131] Hrimiuc, p.298

[132] (Romanian) Andrei Stavilă, "Eveniment: Jurnalul lui Petru Comarnescu", in *Convorbiri Literare*, January 2005

[133] Hrimiuc, p.292, 302

[134] Hrimiuc, p.292-293, 295

[135] Cernat (2007), p.152; Ghemeș, p.73

[136] Ghemeș, p.73-75

[137] Gabriel Liiceanu, *The Păltiniș Diary: A Paideic Model in Humanist Culture*, Central European University Press, Budapest & New York City, 2000, p.22-23. ISBN 963-9116-89-0

[138] (Romanian) Horia Gârbea, "Locuri de taină și desfrîu", in *România Literară*, Nr. 49/2008

[139] (Romanian) Dumitru Radu Popa, "Între două povețe: spiritul exaltat și spiritul treaz", in *Viața Românească*, Nr. 1-2/2007, p.33

[140] (Romanian) Cosmin Ciotloș, "Masca transparentă", in *România Literară*, Nr. 20/2010

[141] Teodoreanu & Ruja, p.10, 14

[142] (Romanian) Ion Bălu, "Prezența discretă a lui George Lesnea", in *Convorbiri Literare*, April 2002

[143] Popa, p.91, 93

[144] Teodoreanu & Ruja, p.15-16

[145] Pîrjol, p.19, 25; Teodoreanu & Ruja, p.16

[146] (Romanian) Gina Popa, "Academia Liberă 'Păstorel' aniversează 37 de ani", in *Evenimentul*, February 7, 2012

[147] Teodoreanu & Ruja, p.8, 16

24.5 References

• Păstorel Teodoreanu, Alexandru Ruja, *Tămâie și otravă*, Editura de Vest, Timișoara, 1994. ISBN 973-36-0165-9

• Lucian Boia, *Capcanele istoriei. Elita intelectuală românească între 1930 și 1950*, Humanitas, Bucharest, 2012. ISBN 978-973-50-3533-4

• George Călinescu, *Istoria literaturii române de la origini pînă în prezent*, Editura Minerva, Bucharest, 1986

• Paul Cernat, *Avangarda românească și complexul periferiei: primul val*, Cartea Românească, Bucharest, 2007. ISBN 978-973-23-1911-6

• Ileana Ghemeș, "Drumul revistei *Țara Noastră* în 1925", in the December 1 University of Alba Iulia *Philologica Yearbook*, 2002, p. 66-75

• Gheorghe Hrimiuc, postface and notes to Al. O. Teodoreanu, *Hronicul Măscăriciului Vălătuc*, Editura Junimea, Iași, 1989, p. 292-334. ISBN 973-37-0003-7

• Eugen Lovinescu, *Istoria literaturii române contemporane*, Editura Minerva, Bucharest, 1989. ISBN 973-21-0159-8

• (Romanian) Doris Mironescu, "*Craii* lui Păstorel. De la *savoir vivre* la *savoir mourir*", in *Timpul*, Nr. 9/2008, p. 16-17

• (Romanian) Florina Pîrjol, "Destinul unui formator de gusturi. De la savoarea 'pastilei' gastronomice la gustul fad al compromisului", in *Transilvania*, Nr. 12/2011, p. 16-26

• Alexandru Piru, *Viața lui G. Ibrăileanu*, Fundația Regală pentru Literatură și Artă, Bucharest, 1946

• (Romanian) Mircea Popa, "Ștefan Baciu - colaborări și versuri uitate", in *Steaua*, Nr. 10-11, October–November 2011, p. 90-93

24.6 External links

• Excerpt from Al. O. Teodoreanu, *Son of a Bitch*, in the Romanian Cultural Institute *Plural Magazine*, Nr. 26/2005

Chapter 25

Keith Wallace (wine writer)

Keith Wallace, M.S. Oenology and Viticulture (University of California, Davis) is the wine columnist for The Daily Beast[1] and also founded the Wine School of Philadelphia.

Previously he served as an executive chef and a journalist for National Public Radio, as well as a winemaker and wine consultant both in the United States and Italy.[2] He has contributed to Philadelphia Magazine, Philadelphia Style, Barron's New Wine Lovers Companion, among other publications, and also created the Philly Uncorked show for www.philly.com.

His forthcoming food and wine book for Running Press, "Corked & Forked", is scheduled for publication in 2011.

25.1 Controversy

The Wine School received national press attention in 2009 when the WWE challenged its trademark application with the U.S. Patent & Trademark Office for the mark, "Sommelier Smackdown".[3] Litigation surrounding the school's intellectual property rights is ongoing.

Also in 2009, the Wine School and founder Keith Wallace were featured on NPR's "All Things Considered"[4] as a result of Mr. Wallace's controversial article published in The Daily Beast, "How Wine Became Like Fast Food".[5]

25.2 References

[1] http://www.thedailybeast.com/author/keith-wallace/

[2] http://citypaper.net/articles/2007/04/12/class-act

[3] http://www.winespectator.com/webfeature/show/id/41062

[4] http://www.npr.org/templates/story/story.php?storyId=120846050

[5] http://www.thedailybeast.com/blogs-and-stories/2009-11-03/how-wine-became-like-fast-food/

25.3 External links

- The Wine School Website
- The Wine School Blog

25.4 Text and image sources, contributors, and licenses

25.4.1 Text

- **Oenologists** *Source:* https://en.wikipedia.org/wiki/Oenology?oldid=691507404 *Contributors:* Olivier, Dante Alighieri, Stan Shebs, Bogdangiusca, Andres, Wnissen, Lfh, Phoebe, Robbot, Hadal, UtherSRG, T0m, Mike R, Tsemii, Nina Gerlach, SpeedyGonsales, Justinc, MCiura, Saxifrage, Waldir, DePiep, Bgwhite, YurikBot, Red Slash, Splash, Eleassar, Jpbowen, Emijrp, Funkendub, Eskimbot, Amatulic, Libertines, Scharks, SashatoBot, NikoSilver, Good Intentions, Bjankuloski06en~enwiki, Fernando S. Aldado~enwiki, Yms, Meco, Van helsing, MooglePower, MJBoa, Agne27, Thijs!bot, Wikid77, Bobblehead, AgentPeppermint, Natalie Erin, Doremítzwr, Freddiem, Nipisiquit, Deflective, .anacondabot, Kingdom85, Grey Wanderer, Funandtrvl, VolkovBot, TXiKiBoT, AlleborgoBot, SieBot, Wahrmund, Muhends, Kleinhev, Tomas e, Ottawahitech, Viguerie, Dthomsen8, Addbot, RivalPeeps, Download, CarsracBot, AndersBot, Kivar2, Jason Recliner, Esq., Anypodetos, Roumpf, Xqbot, Flaneur628, GrouchoBot, Omnipaedista, RibotBOT, Trafford09, Haeinous, Beaukarpo, GoingBatty, Anglais1, Adrabenco, Insommia, ClueBot NG, -sche, Curb Chain, Northamerica1000, Testem, ColonelHenry, DerekWinters, Giorgi Balakhadze, Rormond, Fluoresce, Ghp71, Cmarchual and Anonymous: 55

- **Maynard Amerine** *Source:* https://en.wikipedia.org/wiki/Maynard_Amerine?oldid=687763972 *Contributors:* Phoebe, Ukexpat, Magioladitis, WWGB, Troutfang, Bamyers99, Helpful Pixie Bot, MrNiceGuy1113, UCDSpecColl, Archivist D, Ranganathanwasright and KasparBot

- **Christine Barbe** *Source:* https://en.wikipedia.org/wiki/Christine_Barbe?oldid=568520661 *Contributors:* Klemen Kocjancic, Agne27, Missvain, GrahamHardy and Tassedethe

- **Ruy Barbosa Popolizio** *Source:* https://en.wikipedia.org/wiki/Ruy_Barbosa_Popolizio?oldid=622750961 *Contributors:* Scanlan, AnomieBOT and EmausBot

- **Valentin Blattner** *Source:* https://en.wikipedia.org/wiki/Valentin_Blattner?oldid=691477294 *Contributors:* Vegaswikian, Agne27, Dr. Blofeld, Niceguyedc, Lantus and Anonymous: 2

- **Yiannis Boutaris** *Source:* https://en.wikipedia.org/wiki/Yiannis_Boutaris?oldid=680545433 *Contributors:* Cplakidas, Scanlan, Daemonic Kangaroo, Paulmcdonald, Lemur12, GirasoleDE, Place Clichy, Addbot, Тиверополник, Luckas-bot, Guy1890, Yeomos, Lapost, Xqbot, Erud, Omnipaedista, Candotti, LucienBOT, HRoestBot, Philly boy92, Trappist the monk, Reaper Eternal, EmausBot, Dolescum, ClueBot NG, Κυρίλλος, BG19bot, Epirotes1, BattyBot, Monkbot, Absolute98 and Anonymous: 8

- **Cathy Corison** *Source:* https://en.wikipedia.org/wiki/Cathy_Corison?oldid=624500108 *Contributors:* Mandarax, Cydebot, Missvain, Tassedethe, Ɱ, RjwilmsiBot and Mdann52

- **Tullio De Rosa** *Source:* https://en.wikipedia.org/wiki/Tullio_De_Rosa?oldid=649964639 *Contributors:* Bearcat, Tabletop, Bgwhite, Joel7687, Chris the speller, GoodDay, Symposiarch, Agne27, DGG, Katharineamy, Yobot, Tiziano.fantuzzi, In ictu oculi, John of Reading, Captain Assassin!, ClueBot NG, TitiusSapientius, Hmainsbot1 and Anonymous: 2

- **Sarah Gott** *Source:* https://en.wikipedia.org/wiki/Sarah_Gott?oldid=595004712 *Contributors:* Missvain, GrahamHardy, Tassedethe and Monkbot

- **Belinda Kemp** *Source:* https://en.wikipedia.org/wiki/Belinda_Kemp?oldid=630093785 *Contributors:* Neko-chan, Woohookitty, Symposiarch, Keith D, Tomas e, Vejvančický, ClueBot NG, BattyBot and Sunny48

- **Mia Klein** *Source:* https://en.wikipedia.org/wiki/Mia_Klein?oldid=639851466 *Contributors:* Cydebot, Agne27, Missvain, Waacstats, Mufka, GrahamHardy, Mercurywoodrose, Tassedethe, Mdann52, Mogism, Jamesmcmahon0, Selene Wines and Mia Klein

- **Max Léglise** *Source:* https://en.wikipedia.org/wiki/Max_L%C3%A9glise?oldid=659435626 *Contributors:* Olivier Debre, John Vandenberg, Ser Amantio di Nicolao, Agne27, Fetchcomms, Waacstats, Nono64, BotMultichill, LilHelpa, RjwilmsiBot, Terfilo, KLBot2, VIAFbot, KasparBot and Anonymous: 1

- **Zelma Long** *Source:* https://en.wikipedia.org/wiki/Zelma_Long?oldid=628985951 *Contributors:* Afasmit, Dr. Blofeld, Mercurywoodrose, Rosiestep, Piledhigheranddeeper, Jandlgilbert, ComputerJA and Bonkers The Clown

- **Justin Meyer** *Source:* https://en.wikipedia.org/wiki/Justin_Meyer?oldid=685006256 *Contributors:* Cydebot, Dr. Blofeld, Connormah, Mercurywoodrose, Ɱ, Blofeld Dr., Bonkers The Clown, Lemnaminor and YouWannaSee

- **Hermann Müller (Thurgau)** *Source:* https://en.wikipedia.org/wiki/Hermann_M%C3%BCller_(Thurgau)?oldid=685624620 *Contributors:* Magnus Manske, Pablo X, Stemonitis, Chroniclev, Emijrp, SmackBot, DantheCowMan, Ser Amantio di Nicolao, Svartkell, Symposiarch, Chicheley, Cydebot, Agne27, Thijs!bot, Esowteric, JustAGal, Waacstats, Riccardobot, VolkovBot, FlagSteward, Monegasque, Rosiestep, Tomas e, MystBot, Addbot, Roland.h.bueb, Kestrel man hoovers in the dark, LilHelpa, MauritsBot, DSisyphBot, XZeroBot, RjwilmsiBot, Spray787, VIAFbot, Cmstammen, JoachimKohlerBremen, KasparBot, MapleSyrupRain and Anonymous: 2

- **List of oenologists** *Source:* https://en.wikipedia.org/wiki/List_of_oenologists?oldid=665918425 *Contributors:* Emijrp, Primecoordinator, Missvain, Dr. Blofeld, Tassedethe, Legobot, Srich32977, Blueberryzoo, JoaquinMayo and Anonymous: 5

- **Ottavio Ottavi** *Source:* https://en.wikipedia.org/wiki/Ottavio_Ottavi?oldid=666043963 *Contributors:* Rjwilmsi, Attilios, Ser Amantio di Nicolao, Ian Spackman, Waacstats, Addbot, RjwilmsiBot, VIAFbot and KasparBot

- **Émile Peynaud** *Source:* https://en.wikipedia.org/wiki/%C3%89mile_Peynaud?oldid=674754934 *Contributors:* Zoicon5, Bearcat, Worc63, Blorg, Bender235, Merope, SlaveToTheWage, Rjwilmsi, Walter Moar, Chanlyn, SmackBot, Symposiarch, Stuart Drewer, Agne27, AJD, Dr. Blofeld, Murgh, Bordeaux2000, BOTijo, Tomas e, Addbot, Koppas, Sisyph, Citation bot, Fieldday-thursday, RjwilmsiBot, ClueBot NG, BattyBot, VIAFbot, Monkbot, KasparBot and Anonymous: 5

- **Jacques Puisais** *Source:* https://en.wikipedia.org/wiki/Jacques_Puisais?oldid=659435605 *Contributors:* Olivier Debre, Ser Amantio di Nicolao, Agne27, Waacstats, Nono64, Yobot, LilHelpa, RjwilmsiBot, VIAFbot and KasparBot

- **Michel Rolland** *Source:* https://en.wikipedia.org/wiki/Michel_Rolland?oldid=691556680 *Contributors:* Bearcat, Robbot, Mboverload, Gadfium, Blorg, Marudubshinki, David Justin, Rlove, Algae, SmackBot, Hetar, CmdrObot, Don Serapio~enwiki, Agne27, Jean Marc, Peter Francisco, Po Boy, Dr. Blofeld, Ericoides, Murgh, Smoulding, Waacstats, Fabrictramp, New Zealand Wine Drinker, Prhartcom, TreasuryTag, AlleborgoBot, Tomas e, Addbot, Hlcollins, Full-date unlinking bot, RjwilmsiBot, John of Reading, VIAFbot, Mandruss, KasparBot and Anonymous: 13

- **Carol Shelton** *Source:* https://en.wikipedia.org/wiki/Carol_Shelton?oldid=673362745 *Contributors:* Klemen Kocjancic, Afasmit, Cydebot, Agne27, Missvain, Tassedethe, RjwilmsiBot, Mdann52, Monkbot, Devon Ward and Anonymous: 2

- **Kilien Stengel** *Source:* https://en.wikipedia.org/wiki/Kilien_Stengel?oldid=685849928 *Contributors:* D6, Naraht, Afasmit, GoodDay, Ohconfucius, Ser Amantio di Nicolao, CmdrObot, Waacstats, SalomonCeb, Monegasque, Plastikspork, Tomas e, Niceguyedc, Hedonistevatel, XLinkBot, Addbot, Lightbot, Yobot, Ptbotgourou, AnomieBOT, Omnipaedista, FrescoBot, Redrose64, Jonesey95, RjwilmsiBot, SporkBot, BG19bot, Mark Arsten, ChrisGualtieri, VIAFbot, KasparBot and Anonymous: 5

- **Lane Tanner** *Source:* https://en.wikipedia.org/wiki/Lane_Tanner?oldid=688584730 *Contributors:* Agne27, Missvain, GrahamHardy, Tassedethe, GoingBatty, Doctree, Mdann52, Monkbot and Anonymous: 1

- **Păstorel Teodoreanu** *Source:* https://en.wikipedia.org/wiki/P%C4%83storel_Teodoreanu?oldid=691010328 *Contributors:* Bogdangiusca, Bgwhite, EamonnPKeane, Dahn, Bejnar, Ser Amantio di Nicolao, Cydebot, Biruitorul, R'n'B, Tgeairn, Mycomp, Yngvadottir, Muleiolenimi, EmausBot, Mogism and Anonymous: 5

- **Keith Wallace (wine writer)** *Source:* https://en.wikipedia.org/wiki/Keith_Wallace_(wine_writer)?oldid=665918208 *Contributors:* DaGizza, SmackBot, Agne27, Johnpacklambert, TreasuryTag, Malcolmxl5, Tomas e, Auntof6, Tassedethe, Yobot, DynamoDegsy, Vino Magister, Decathlete and Anonymous: 1

25.4.2   Images

- **File:1950_Cooperativă_de_consum._Secţia_restaurant.jpg** *Source:*   https://upload.wikimedia.org/wikipedia/commons/c/c9/1950_Cooperativ%C4%83_de_consum._Sec%C5%A3ia_restaurant.jpg *License:* Public domain *Contributors:* http://www.comunismulinromania.ro/Arhiva-foto/Cotidian/Cotidian-I.html *Original artist:* unknown/uncredited

- **File:Ambox_important.svg** *Source:* https://upload.wikimedia.org/wikipedia/commons/b/b4/Ambox_important.svg *License:* Public domain *Contributors:* Own work, based off of Image:Ambox scales.svg *Original artist:* Dsmurat (talk · contribs)

- **File:Commons-logo.svg** *Source:* https://upload.wikimedia.org/wikipedia/en/4/4a/Commons-logo.svg *License:* ? *Contributors:* ? *Original artist:* ?

- **File:Corn_01.JPG** *Source:*   https://upload.wikimedia.org/wikipedia/commons/0/03/Corn_01.JPG *License:*   CC BY-SA 3.0 *Contributors:* Transferred from ml.wikipedia by Sreejith K (<a href='//commons.wikimedia.org/wiki/User_talk:Sreejithk2000' title='User talk: Sreejithk2000'>talk</a>) *Original artist:* Original uploaded by Ashlyak.

- **File:Crystal_Clear_app_Login_Manager_2.png** *Source:*   https://upload.wikimedia.org/wikipedia/en/c/c2/Crystal_Clear_app_Login_Manager_2.png *License:* ? *Contributors:* ? *Original artist:* ?

- **File:Flag_of_Chile.svg** *Source:* https://upload.wikimedia.org/wikipedia/commons/7/78/Flag_of_Chile.svg *License:* Public domain *Contributors:* Own work *Original artist:* SKopp

- **File:Flag_of_Italy.svg** *Source:* https://upload.wikimedia.org/wikipedia/en/0/03/Flag_of_Italy.svg *License:* PD *Contributors:* ? *Original artist:* ?

- **File:Glass_of_wine.svg** *Source:* https://upload.wikimedia.org/wikipedia/commons/d/d7/Glass_of_wine.svg *License:* CC-BY-SA-3.0 *Contributors:*

- Glass_of_wine.png *Original artist:* Glass_of_wine.png: Jndrline

- **File:Hermann_Müller_(1850-1927).jpg** *Source:*   https://upload.wikimedia.org/wikipedia/commons/b/b9/Hermann_M%C3%BCller_%281850-1927%29.jpg *License:* Public domain *Contributors:* Hessische Forschungsanstalt für Wein-, Obst- und Gartenbau Geisenheim/Rhein (Hrsg.): *Geisenheim 1872-1972. 100 Jahre Forschung und Lehre für Wein-, Obst- und Gartenbau.* Verlag Eugen Ulmer, Stuttgart 1972 ISBN 3-8001-3023-8 *Original artist:* The original uploader was Martin Bahmann at German Wikipedia

- **File:Justin_Meyer3.jpg** *Source:* https://upload.wikimedia.org/wikipedia/commons/c/cb/Justin_Meyer3.jpg *License:* CC BY-SA 3.0 *Contributors:* Image provided by request from Silver Oak Cellars. *Original artist:* Silver Oak Cellars.

- **File:Justin_Meyer4.jpg** *Source:* https://upload.wikimedia.org/wikipedia/commons/4/45/Justin_Meyer4.jpg *License:* CC BY-SA 3.0 *Contributors:* Image provided by request from Silver Oak Cellars. *Original artist:* Silver Oak Cellars.

- **File:Justin_Meyer_and_Ray_Duncan.jpg** *Source:*   https://upload.wikimedia.org/wikipedia/commons/6/67/Justin_Meyer_and_Ray_Duncan.jpg *License:* CC BY-SA 3.0 *Contributors:* Image provided by request from Silver Oak Cellars. *Original artist:* Silver Oak Cellars.

- **File:L'indétronâble_Michel_Rolland.jpg** *Source:* https://upload.wikimedia.org/wikipedia/commons/3/36/L%27ind%C3%A9tron%C3%A2ble_Michel_Rolland.jpg *License:* CC BY 2.0 *Contributors:* originally posted to **Flickr** as L'indétronâble Michel Rolland *Original artist:* Ludovic Roif

- **File:Leaf_1_web.jpg** *Source:* https://upload.wikimedia.org/wikipedia/commons/f/f4/Leaf_1_web.jpg *License:* Public domain *Contributors:* PdPhoto *Original artist:* Jon Sullivan

- **File:Michel_Rolland'{}s_Vineyard_of_Château_Le_Bon_Pasteur,_Pomerol.jpg** *Source:*   https://upload.wikimedia.org/wikipedia/commons/d/d6/Michel_Rolland%27s_Vineyard_of_Ch%C3%A2teau_Le_Bon_Pasteur%2C_Pomerol.jpg *License:* CC BY 2.0 *Contributors:* Flickr: Château Le Bon Pasteur, Pomerol *Original artist:* Megan Mallen

25.4.3 Content license

www.ingramcontent.com/pod-product-compliance
Lightning Source LLC
Chambersburg PA
CBHW080647180526
45168CB00008B/3335